T0134426

Advanced Sciences and Technologies for Security Applications

Indexed by SCOPUS

The series Advanced Sciences and Technologies for Security Applications comprises interdisciplinary research covering the theory, foundations and domain-specific topics pertaining to security. Publications within the series are peer-reviewed monographs and edited works in the areas of:

- biological and chemical threat recognition and detection (e.g., biosensors, aerosols, forensics)
- crisis and disaster management
- terrorism
- cyber security and secure information systems (e.g., encryption, optical and photonic systems)
- traditional and non-traditional security
- energy, food and resource security
- economic security and securitization (including associated infrastructures)
- transnational crime
- human security and health security
- social, political and psychological aspects of security
- recognition and identification (e.g., optical imaging, biometrics, authentication and verification)
- smart surveillance systems
- applications of theoretical frameworks and methodologies (e.g., grounded theory, complexity, network sciences, modelling and simulation)

Together, the high-quality contributions to this series provide a cross-disciplinary overview of forefront research endeavours aiming to make the world a safer place.

The editors encourage prospective authors to correspond with them in advance of submitting a manuscript. Submission of manuscripts should be made to the Editor-in-Chief or one of the Editors.

More information about this series at http://www.springer.com/series/5540

Anthony J. Masys
Editor

Sensemaking for Security

 Springer

Editor
Anthony J. Masys
College of Public Health
University of South Florida
Tampa, FL, USA

ISSN 1613-5113 ISSN 2363-9466 (electronic)
Advanced Sciences and Technologies for Security Applications
ISBN 978-3-030-72000-1 ISBN 978-3-030-71998-2 (eBook)
https://doi.org/10.1007/978-3-030-71998-2

This Springer imprint is published by the registered company Springer Nature Switzerland AG
The registered company address is: Gewerbestrasse 11, 6330 Cham, Switzerland

Contents

The Security Landscape—Systemic Risks Shaping Non-traditional Security

Anthony J. Masys

Abstract Today's security landscape is being shaped by an inherent Volatility, Uncertainty, Complexity, Ambiguity (VUCA). The threats and risks to safety and security derive from both man-made and natural circumstances. Events like Hurricane Michael (2018), Hurricane Maria (2017), COVID-19 pandemic (2020), Ebola Outbreak (2014–2016), Hurricane Katrina (2005), Fukushima Daiichi nuclear meltdown (2011), Typhoon Haiyan (2013) and global terrorist events illustrate the devastating effects of natural and man-made disasters on human systems and human security (Masys et al. in Procedia Econ Financ 18:772–779, 2014 [1]). As reported in Masys and Lin (Asia/Pacific security challenges—managing black swans and persistent threats. Springer Publishing, 2017 [2]), over the past four decades there have been a growing number of small and medium-scale disasters which have resulted in a total loss of over US$1.15 trillion. This threat and risk landscape challenges regional security along such lines as: national security; energy security; water security; food security; health security; human security; environmental security; economic security. Considering the current COVID-19 pandemic, as of November 2020, the COVID-19 pandemic has seen upwards of 60 million cases, 1.4 million deaths, and has significantly impacted the global economy and affected the most vulnerable. The impacts have seen the closure of borders, economic disruptions and failures, strained and overwhelmed health care systems, failure of supply chains all of which are contributing to a human and national security issue. The COVID 19 represents a national and global health security disaster revealing systemic vulnerabilities. A disease anywhere is a disease everywhere. Threats to global human security reside within disease vectors that are transnational making sensemaking an essential element of security. Sensing the threat and risk landscape that characterizes the security environment requires an understanding of the inherent systems, interdependencies, nonlinearity across the security domain. This chapter provides an overview of the complex security landscape and the inherent systemic risks that shape the security discourse.

Keywords Security · VUCA · COVID-19 · Nontraditional security

A. J. Masys (✉)
College of Public Health, University of South Florida, Tampa, FL, USA

International Centre for Policing and Security, University of South Wales, Cardiff, UK

© The Author(s), under exclusive license to Springer Nature Switzerland AG 2021
A. J. Masys (ed.), *Sensemaking for Security*, Advanced Sciences and Technologies
for Security Applications, https://doi.org/10.1007/978-3-030-71998-2_1

1 Introduction

The threats and risks to safety and security derive from both man-made and natural circumstances. Events like Hurricane Michael (2018), Hurricane Maria (2017), COVID-19 pandemic (2020), Ebola Outbreak (2014–2016), Hurricane Katrina (2005), Fukushima Daiichi nuclear meltdown (2011), Typhoon Haiyan (2013) and global terrorist events illustrate the devastating effects of natural and man-made disasters on human systems and human security [1]. As reported in Masys and Lin [2], over the past four decades there have been a growing number of small and medium-scale disasters which have resulted in a total loss of over US$1.15 trillion. This threat and risk landscape challenges regional security along such lines as: national security, energy security; water security; food security; health security; human security; environmental security; economic security. Considering the current COVID-19 pandemic, as of November 2020, the COVID-19 pandemic has seen upwards of 60 million cases, 1.4 million deaths, and has significantly impacted the global economy and affected the most vulnerable.

The impacts of such events cannot be seen in isolation. WEF Global Risks reports have clearly articulated the hyperconnected risks that permeate the world and reflect the inherent '… fragility and vulnerabilities that lie within the social/technological/economic/political/ecological interdependent systems' [1]. It is through these underlying networks that Helbing [3, p. 51] argues that we have '… created pathways along which dangerous and damaging events can spread rapidly and globally' and thereby has increased systemic risks.

A Chatham House report, 'Preparing for High Impact, Low Probability Events', found that governments and businesses remain unprepared for such events [4]. Helbing [3, p. 51] argues that:

> Many disasters in anthropogenic systems should not be seen as 'bad luck', but as the results of inappropriate interactions and institutional settings. Even worse, they are often the consequences of a wrong understanding due to the counter-intuitive nature of the underlying system behaviour.

Today's security environment is being shaped by an inherent Volatility, Uncertainty, Complexity, Ambiguity (VUCA). As described in Kraaijenbrink [5], the components are defined as:

Volatility—Volatility refers to the speed of change in an industry, market or the world in general. It is associated with fluctuations in demand, turbulence and short time to markets and it is well-documented in the literature on industry dynamism. The more volatile the world is, the more and faster things change.

Uncertainty—Uncertainty refers to the extent to which we can confidently predict the future. Part of uncertainty is perceived and associated with people's inability to understand what is going on. Uncertainty, though, is also a more objective characteristic of an environment. Truly uncertain environments are those that don't allow any prediction, also not on a statistical basis. The more uncertain the world is, the harder it is to predict.

Complexity—Complexity refers to the number of factors that we need to take into account, their variety and the relationships between them. The more factors, the greater their variety and the more they are interconnected, the more complex an environment is. Under high complexity, it is impossible to fully analyze the environment and come to rational conclusions. The more complex the world is, the harder it is to analyze.

Ambiguity—Ambiguity refers to a lack of clarity about how to interpret something. A situation is ambiguous, for example, when information is incomplete, contradicting or too inaccurate to draw clear conclusions. More generally it refers to fuzziness and vagueness in ideas and terminology. The more ambiguous the world is, the harder it is to interpret.

These four characteristics are shaping the security landscape in terms of not only events but also with regards to our understanding and sensemaking of security events and implications. Decision making under such VUCA conditions becomes problematic due to the unintended consequences that may emerge from a lack of system understanding. This becomes a challenge for crisis leadership.

When we consider the security landscape under VUCA conditions, the words of Woods [6, p. 316] certainly resonate. He asks the question:

> How do people detect that problems are emerging or changing when information is subtle, fragmented, incomplete or distributed across different groups involved in production processes and in safety management. Many studies have shown how decision makers in evolving situations can get stuck in a single problem frame and miss or misinterpret new information that should force re-evaluation and revision of the situation assessment …

There is an urgent need to re-examine our exposure to security risks and vulnerabilities, in order to transform our understanding and perception of security. This essentially questions how we **make sense** of global security risks, vulnerabilities and impacts. Key is understanding the security landscape in terms of inherent systemic risks. Local and regional security and public safety events derived from natural or manmade disasters can precipitate systemic risks that reach global proportions. For example the 2011 Thailand flooding had consequences on global supply chains. Similarly, as described in Masys et al. [1]:

> With events such as the 2010 Ash Cloud stemming from the eruption of Eyjafjallajökull and the resulting disruptions to air travel and trade in Europe [7], we see how 'networked risks' are not confined to national borders or a single sector, and do not fit the monocausal model of risk. As argued by Renn et al. [8, p. 234] such risks or hyper-risks are '… complex (multi-causal) and surrounded by uncertainty and/or ambiguity'.

Systems thinking emerges as a powerful lens in which to support sensemaking. Normal accident theory [9], latent risks, resident pathogens [10], normalization of deviance [11], incubation [12] reflect how the seeds of disasters and crisis reside within the very societal systems. Systems thinking provides a lens in which to unearth the interconnectivity and interdependencies in complex societal systems. As noted in OECD [13, p. 14], 'The excuse that dangers are clear only in hindsight does not stand up to objective scrutiny. Major simulation exercises in OECD countries predicted accurately how a crisis like Covid-19 could unfold, but they were not acted on, or not sufficiently, judging by what has happened'.

Systemic interdependence leads to relative opacity when it comes to understanding the propagation of risk and how it shapes the security landscape. Like natural **ecosystems**, the **security ecosystem** comprises a variety of diverse actors (human, physical and informational) as described in Masys and Lin [2] that interact in complex and dynamic ways (Fig. 1).

Systemic risks are defined in Centeno et al. [14, p. 68]:

> … as the threat that individual failures, accidents, or disruptions present to a system through the process of contagion. This is the risk that unexpected and unlikely interactions may lead to unpredicted threats to system survival. We define emerging risk as the danger that arises from new technologies or interdependencies—disasters that we have not experienced yet are expected to increase in both frequency and magnitude, such as terrorist attacks or natural disasters.

Within this networked and interdependent security landscape, cascading risk is expected with the growing interconnectedness of natural and man-made systems. As noted with the Thailand flooding (2011) and the Ash Cloud of (2010), individual events, even localized ones, may cause large repercussions globally.

Systemic risk emerges from the interconnectedness of the networks underlying a system and result from the interactions of individual risks resulting in cascades of secondary and tertiary failures. The interconnectedness associated with systems

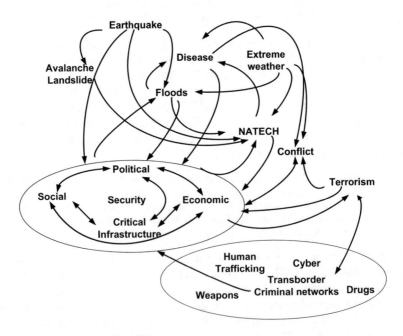

Fig. 1 Security ecosystem [2, p. 11]

Table 1 Top 5 risks [23]

Top 5 global risks in terms of likelihood	Top 5 global risks in terms of impact
Extreme weather	Climate action failure
Climate action failure	Weapons of mass destruction
Natural disasters	Biodiversity loss
Biodiversity loss	Extreme weather
Human-made environmental disasters	Water crises

makes predicting the probability, cost and magnitude of risk difficult. As such, anticipating the distribution of impacts is problematic often reifying 'glocally' (with both local and global impact). A localized drought in one part of the world can raise food prices globally and result in food insecurity in other regions often impacting the most vulnerable populations. This can result in geopolitical instabilities further exacerbating disruptions locally, regionally and globally. When we consider risk mitigation through the lens of systemic risk, we recognize that traditional models of local mitigation are ill suited for such cascading risks.

The COVID-19 pandemic is a perfect example. It is often phrased 'a disease anywhere is a disease everywhere'. Centeno et al. [14, p. 77] argues that 'this same global interconnectedness that supports trade and global supply chains, has also increased the potential for catastrophic effects from the spread of zoonoses, i.e., diseases transmitted from animals to humans'. This represents an inherent structural fragility in our global systems. The distribution of impacts is often not equitable, affecting the most vulnerable stemming from the inherent global inequality between rich and poor'. Renn et al. [15, p. 402] argue that '… risk society is a catastrophic society' [16, p. 24] … if we look at globally interconnected, non-linear risks such as those posed by climate change [17], the present global financial system [18], geopolitics in the nuclear age [19], pandemics [20], or the growing inequality between rich and poor [21]'. Boin et al. [22] refers to COVID-19 as a 'creeping crisis' highlighting the resident pathogens and latent risks that reside within our societal infrastructure.

2 Systemic Risk and the WEF Risk Landscape (2020)

The risk landscape presented by the WEF [23] report illustrates the interconnectivity, interdependency and complexity associated with risks across the following domains: Economic, Environmental, Geopolitical, Societal, and Technological. The top 5 risks [23] associated with likelihood and impacts are:

As depicted in Table 1 and Fig. 1, these risks shape the security ecosystem and points to embracing a perspective that moves beyond traditional security to one that recognizes nontraditional security concerns.

3 Traditional Security and Nontraditional Security

The realization of a new security calculus has emerged. One that recognizes threats other than war and military operations has become a focal point of national security strategies. Stares [24, p. 128] describes this new security agenda in terms of '… conception of "human" or "global" security is tantamount, however, to accepting a still broader set of problems as security concerns: poverty, political injustice, natural disasters, crime, social discrimination, and unemployment, to name just some. In effect, these items are not merely additions to the list of traditional concerns, rather, they are the new security agenda'. Singh and Nunes [25, p. 109] define non-traditional security issues as '… as non-military threats that threaten either the political and social integrity of a nation-state or the health of its inhabitants. Such threats can be from within or across the borders, which can also be termed as low-intensity conflicts'.

As noted in Wong and Brown [26, p. 80], 'Nontraditional Security is a term which has been adopted in both policy and academic spheres to describe non-military threats. The range of possible sources of nontraditional security threats is vast, encompassing threats emanating from agents (e.g. terrorists), events (e.g. natural disasters) and processes (e.g. climate change). Nontraditional security threats, by their nature, do not respect borders and therefore it is impossible for any individual actor or state to develop a comprehensive solution autonomously'.

4 Human Security

Such threats to security are becoming complex and multifaceted and are increasingly challenging traditional notions of security. As detailed in UNOCHA [27], the need for a new paradigm of security is associated with two sets of dynamics:

- First, human security is needed in response to the complexity and the interrelatedness of both old and new security threats—from chronic and persistent poverty to ethnic violence, human trafficking, climate change, health pandemics, international terrorism, and sudden economic and financial downturns. Such threats tend to acquire transnational dimensions and move beyond traditional notions of security that focus on external military aggressions alone.
- Second, human security is required as a comprehensive approach that utilizes the wide range of new opportunities to tackle such threats in an integrated manner. Human security threats cannot be tackled through conventional mechanisms alone. Instead, they require a new consensus that acknowledges the linkages and the interdependencies between development, human rights and national security.

This certainly characterizes a nontraditional security challenge/threat capturing the complex interdependencies of the system.

Challenging the traditional notion of security, the articulation of human security represents a central referent for nontraditional security. The relevance of the human security construct within the context of the current COVID-19 pandemic certainly shines through. The Commission of Human Security efforts reconceptualized security by:

i. moving away from traditional, state-centric conceptions of security that focused primarily on the safety of states from military aggression, to one that concentrates on the security of the individuals, their protection and empowerment;

ii. drawing attention to a multitude of threats that cut across different aspects of human life and thus highlighting the interface between security, development and human rights; and

iii. promoting a new integrated, coordinated and people-centered approach to advancing peace, security and development within and across nations [27] (Table 2).

The increased transnational flow of people, goods, money and information as products of 'globalization' has also shaped the security landscape. The transnational/transborder nature of security challenges our traditional views of security characterized by state-based, military dimensions. The nontraditional security calculus thereby emerges as part of the security landscape that can often have significant national security implications. Dimensions of the security calculus such as health security, economic security, food security and energy security are interrelated concepts that characterize the security landscape as complex. Actions and interventions associated with this complex problem space thus can have highly unpredictable and unintended consequences.

Issues such as climate change, resource scarcity, infectious diseases, natural disasters, irregular migration, drug trafficking, information security and transnational crime impinge upon the security landscape. As noted in the WEF reports, these

Table 2 Possible types of human security threats [27]

Type of security	Examples of main threats
Economic security	Persistent poverty, unemployment
Food security	Hunger, famine
Health security	Deadly infectious diseases, unsafe food, malnutrition, lack of access to basic health care
Environmental security	Environmental degradation, resource depletion, natural disasters, pollution
Personal security	Physical violence, crime, terrorism, domestic violence, child labor
Community security	Inter-ethnic, religious and other identity based tensions
Political security	Political repression, human rights abuses

global systemic risks move beyond the traditional security challenges associated with national political and territorial integrity (although these threats can influence them), to a more nontraditional security impact that challenges economic and commercial market, as well as act as threat 'triggers' that can initiate a cascading security threat (systemic risks) within a nation and across borders. For example, extreme weather events can create or exacerbate political instability and violence and create destabilizing effects causing mass migration.

Climate change and extreme weather events for example are threatening human security and is considered as one of the biggest non-traditional security threats to national security. Climate change risk has numerous ways to impact the human security. For example climate change and extreme weather events can contribute to food insecurity, water scarcity, floods, internal migrations or displacements, livelihood depletions, resulting in a public health crisis.

For example as noted in Bell and Masys [28],

> Natural hazards have displaced over 26.4 million people per year since 2008. This number will only increase as climate change continues to exacerbate displacement due to monsoon-related flooding, coastal erosion, cyclones, and salinity intrusion. Extreme weather events such as Typhoon Haiyan, Hurricane Maria, and Hurricane Irma shed light on regional and global health stressors that certainly impact global health security. For example, the impact of such events on vulnerable populations in the South Pacific, whereby the effects have created what some might call 'climate refugees'.

> Considering systemic risks associated nontraditional security, the vulnerabilities in regions and communities come into play. These vulnerabilities are shaped by physical, social, economic, and environmental factors and processes that increase the susceptibility of a community/system to the impact of hazards.

With regards to infectious disease and global health security, the WEF [23] report argues that:

> Health systems around the world are at risk of becoming unfit for purpose. Changing societal, environmental, demographic and technological patterns are straining their capacity [23, p. 73]. Considerable progress has been made since the Ebola epidemic in West Africa in 2014–2016, but health systems worldwide are still under-prepared for significant outbreaks of other emerging infectious diseases, such as SARS, Zika and MERS. [23, p. 76]

With regards to COVID-19 impact as a nontraditional security threat, the CFR report by Burwell et al. [29, p. 20] argues that:

> Pandemics are not random events. Pandemics afflict societies through the established relationships that people have created with the environment, other animal species, and each other. The precise timing and location of the coronavirus outbreak that led to this pandemic were difficult to predict, but the emergence of a novel respiratory virus and the threat it would pose to urbanized nations with extensive travel links and underfunded public health systems were not.

5 Cascading Risks

Complex security challenges within a complex risk network structure and behavior (such as the global health security issues associated with the COVID-19 pandemic) can stress test societal systems and reveal 'latent risk' such as the lack of preparation, insufficient vulnerability analysis and response. Weick and Sutcliffe [30, p. 2] highlight how such an event can be '… considered as an abrupt and brutal audit: at a moment's notice, everything that was left unprepared becomes a complex problem, and every weakness comes rushing to the forefront'. Burwell et al. [29] argues that:

> COVID-19 has confirmed the global vulnerabilities that were repeatedly identified in high-level reports, commissions, and intelligence assessments on pandemic threats for nearly two decades prior to this pandemic. COVID-19 has underscored several truths about pandemics and revealed important shortcomings in current global and national capacities to prepare for, detect, and respond to them.

We can reflect upon the sensemaking associated with COVID-19 pandemic in terms of a black swan event [31] or the 'elephant in the room'. As a global health security issue, COVID-19 impacts cross the spectrum of nontraditional security matters. One of the primary features underlying the nontraditional security is the inherent global interdependency and interconnectivity. Helbing [3, p. 51] argues that: '… When analysing today's environmental, health and financial systems or our supply chains and information and communication systems, one finds that these systems have become vulnerable on a planetary scale. They are challenged by the disruptive influences of global warming, disease outbreaks, food (distribution) shortages, financial crashes, heavy solar storms, organized (cyber-) crime, or cyberwar. Our world is already facing some of the consequences: global problems such as fiscal and economic crises, global migration, and an explosive mix of incompatible interests and cultures, coming along with social unrests, international and civil wars, and global terrorism'.

Regarding sensemaking, the potential emergence and systemic risks associated with pandemics has been on the books for years, voiced by foresight and public health experts. As a global security weak signal, COVID-19 can be considered a 'canary in the coal mine'[1] pointing to the fragility of the nontraditional security landscape and creating a human security crisis disrupting global economies and global health.

6 Discussion

Analysis of vulnerabilities associated with nontraditional security requires the discipline of the systems approach which embraces a shift of mind: 'in seeing interrelationships rather than linear cause-effect chains and seeing processes of change rather than snapshots' [32]. Systems thinking thereby is an appropriate approach for communicating such complexities and interdependencies. The conceptual model suggests that there exists a disproportionality of 'causes and effects', in which as Urry

[33, p. 59] remarks, past events are never 'forgotten', they are seeded (manufactured) in the actor network [1, 34].

This is about complexity framing. Hence the inherent interdependencies and inter-connectivity that characterizes the risk space leads to a network model. The notion of hyper-risks [1, 3] captures well the interconnectivity and complexity of the security threats. The complexity lens thereby becomes prominent in examining security and represents a key lens supporting sensemaking.

The lessons learned from such terrorist events as 9/11 and 7/7, Ebola outbreak in Africa, the global financial crisis, have shaped the safety and security agenda. As articulated in the 9/11 report '… failure of imagination and a mindset that dismissed possibilities' point to the requirement for an anticipatory stance regarding disasters and catastrophes. The risk discourse thereby expects analysis to support 'anticipation' of disaster and crisis events. As discussed in Masys [35], the butterfly defect [36], draws attention to the new nature of systemic risk well articulated in the WEF reports. From this we can recognize that small perturbations can have much greater effects and permeate all dimensions of society. Global structural and capability defects in terms of governance and sensemaking means that perturbations can now propagate new forms of risk. Managing such systemic risks is complex. As articulated in Goldin and Mariathasan [36, p. 3]:

> Although each of our individual actions may be rational, collectively, they may lead to failure. … as complexity and integration grow, attribution becomes more difficult and the unintended or unknown consequences of actions increase. Failure to understand or even acknowledge the nonlinear and highly complex nature of global linkages on every level of governance leads to growing weaknesses and can paralyze decision making.

Further, Goldin and Mariathasan [36, p. 66] argue:

> Because we did not understand the systemic vulnerabilities caused by increasing complexity, regulatory arbitrage grew outside the control of regulators. We now know that nodes of the financial network cannot be analyzed in an additive or linear manner … Systemic analysis must examine nodes, pathways, and the relationships between them, because catastrophic changes in the overall state of the system can ultimately derive from how it is organized- from feedback mechanisms within it and from linkages that are latent and often unrecognized.

The systemic risks influencing the human security landscape points to the work of Tavanti and Stachowicz-Stanusch [37]. This can be framed in terms of sensemaking to support crisis and disaster management practices (discussed in detail in this volume Masys 2021). Figure 2 depicts the human security dimensions and its connection to nontraditional security threats and risks.

Given the complexity of the security landscape, characterized by a high degree of volatility, uncertainty, complexity and ambiguity (VUCA), sensemaking will require an anticipatory approach that leverages strategic foresight to explore the systemic risks and emerging vulnerabilities. Scenarios provide insights to develop new ideas to inform more robust decision-making. COVID 19 pandemic is a global threat that impacts global health security, human security, economic security and national security. Lessons from the COVID-19 pandemic pertaining to sensemaking in the security context [23]:

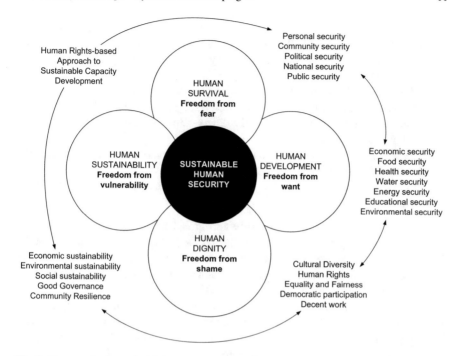

Fig. 2 Framework of sustainable human security (modified from Tavanti and Stachowicz-Stanusch [37, p. 8])

- The risk of a global pandemic has risen in recent years, but the perception of that risk has not matched reality.
- We must be aware of our blind spots and position the world to be ready to react to risks as they occur.

The scientific community foresaw how a 'pandemic would strain the elements of the global supply chains and demands, thereby igniting a cross-border economic disaster because of the highly interconnected world we now live in. By all accounts, the emerging havoc wrought by the pandemic exceeded the predictions in those commentaries' [38].

Connecting the impact of COVID-19 pandemic to the quote from Weick and Sutcliffe [30, p. 2] '... everything that was left unprepared becomes a complex problem, and every weakness comes rushing to the forefront', a number of lessons learned have emerged. The Global Health Security Index which focuses on assessing global health security capacities reported [39] the following findings:

1. National healh security is fundamentally weak around the world. No country is fully prepared for epidemics or pandemics, and every country has important gaps to address.
2. Countries are not prepared for a globally catastrophic biological event.

3. There is little evidence that most countries have tested important health security capacities or shown that they would be functional in a crisis.
4. Most countries have not allocated funding from national budgets to fill identified preparedness gaps.
5. More than half of countries face major political and security risks that could undermine national capability to counter biological threats.
6. Most countries lack foundational health systems capacities vital for epidemic and pandemic response.
7. Coordination and training are inadequate among veterinary, wildlife, and public health professionals and policymakers.
8. Improving country compliance with international health and security norms is essential.

The pandemic (as a national security and global health security issue) reveals the inherent interdependencies, complexity and inequality within the global system. The pandemic represents the reification of underlying systemic risks as detailed in WEF reports. Like COVID-19, exposure to such security events is mediated through a global system that impacts human security (Fig. 1).

Risks shaping the security landscape have become systemic. Treating risks in siloed approaches tends to make weak signals unseen and solutions ineffective. Through a lens of nontraditional security and systemic risks [23], sensemaking can be more strategic revealing opportunities for integrating strategic foresight, scenario planning and transformational resilience. Given the current security landscape and shocks to human systems characterized by complexity and wickedness [35], Goldin and Mariathasan [36, p. 208] argue that 'physical, virtual and social networks need to be constructed in ways that allow them to withstand, and respond to the novel challenges of our time. They have to be flexible and organic rather than static and their capacities cannot be stretched to the limit'. COVID-19 stress tested societal systems and revealed inherent systemic vulnerabilities. The question becomes how do we conceptualize and manage security in the face of extreme events and black swans [31, 34]? Sensemaking thereby emerges as a key enabler in managing systemic risks.

7 Conclusion

Considering the current COVID-19 pandemic, the impacts have seen the closure of borders, economic disruptions and failures, strained and overwhelmed health care systems, failure of supply chains all of which are contributing to a human and national security issue. As noted in Hiscott et al. [40, p. 1], 'despite the shock, there were plenty of warning signs. Since the beginning of the 21st century, recurring outbreaks and epidemics presaged what was coming—there was the first SARS outbreak in 2003, H1N1 influenza pandemic in 2009, MERS coronavirus in 2011, Ebola in 2014–16, mosquito-borne Zika in 2016'.

The COVID-19 represents a threat revealing inherent systemic risks and vulnerabilities throughout our global systems. It has highlighted the nontraditional security lens in which to view the new security discourse. Sensing the threat and risk landscape that characterizes the security environment requires an understanding of the inherent systems, interdependencies, nonlinearity across the security domain. Sensemaking is critical to support crisis management and leadership.

Note

1. https://www.defenseone.com/ideas/2020/05/miners-canary-covid-19-and-rise-non-traditional-security-threats/165446/.

References

1. Masys AJ, Ray-Bennett N, Shiroshita H, Jackson P (2014) High impact/low frequency extreme events: enabling reflection and resilience in a hyper-connected world. In: 4th international conference on building resilience, Salford Quays, 8–11 Sept 2014. Procedia Econ Financ 18:772–779
2. Masys AJ, Lin L (eds) (2018) Asia/Pacific security challenges—managing black swans and persistent threats. Springer Publishing
3. Helbing D (2013) Globally networked risks and how to respond. Nature 497:51–59
4. Lee B, Preston F, Green G (2012) Preparing for high-impact, low-probability events lessons from Eyjafjallajökull- A Chatham House Report. https://www.chathamhouse.org/sites/def ault/files/public/Research/Energy,%20Environment%20and%20Development/r0112_highim pact.pdf
5. Kraaijenbrink J (2018) What does VUCA really mean? Forbes. https://www.forbes.com/sites/ jeroenkraaijenbrink/2018/12/19/what-does-vuca-really-mean/?sh=6fdda5c917d6
6. Woods DD (2006) How to design a safety organization: test case for resilience engineering. In: Hollnagel E, Woods DD, Leveson N (eds) Resilience engineering: concepts and precepts. Ashgate Publishing, Hampshire
7. Harris AJL, Gurioli L, Hughes EE, Lagreulet S (2012) Impact of the Eyjafjallajökull ash cloud: a newspaper perspective. J Geophys Res 117(B00C08):1–35
8. Renn O, Klinke A, van Asselt M (2011) Coping with complexity, uncertainty and ambiguity in risk governance: a synthesis. Ambio 40:231–246
9. Perrow C (1984) Normal accidents: living with high-risk technologies. Basic Books Inc., New York, NY
10. Reason J (1997) Managing the risk of organizational accidents. Ashgate Publishing, Aldershot
11. Vaughan D (1996) The challenger launch decision: risky technology, culture and deviance at NASA. Chicago University Press, London
12. Turner BA (1978) Man-made disasters. Wykeham, London
13. OECD (2020) A systemic resilience approach to dealing with Covid-19 and future shocks. In: New approaches to economic challenges (NAEC), 28 Apr 2020
14. Centeno MA, Nag M, Patterson TS, Shaver A, Windawi AJ (2015) The emergence of global systemic risk. Annu Rev Sociol 41:65–85
15. Renn O, Lucasa K, Haas A, Jaegera C (2019) Things are different today: the challenge of global systemic risks. J Risk Res 22(4):401–415
16. Beck U (1992) Risk society, towards a new modernity. Sage, London
17. IPCC (2014) Climate change 2014: synthesis report. IPCC, Geneva
18. Lo AW (2012) Reading about the financial crisis: a twenty-one-book review. J Econ Lit 50(1):151–178

19. Mearsheimer JJ (2010) The gathering storm: China's challenge to US power in Asia. Chin J Int Politics 3:381–396
20. Lee K (2003) Globalization and health: an introduction. Palgrave Macmillan, London
21. WEF (2017) The global risk report 2017. Accessed 20 July 2017. https://reports.weforum.org/global-risks-2017
22. Boin A, Ekengren M, Rhinard M (2020) Hiding in plain sight: conceptualizing the creeping crisis. Risk Hazards Crisis Public Policy 11(2)
23. WEF (2020) The global risk report 2020. https://www3.weforum.org/docs/WEF_Global_Risk_Report_2020.pdf
24. Stares PB (ed) (1998) New security agenda: a global survey. Japan Center for International Exchange, Tokyo
25. Singh NK, Nunes W (2016) Nontraditional security: redefining state-centric outlook. Jadavpur J Int Relat 20(1):102–124
26. Wong R, Brown S (2016) Stepping up EU-ASEAN cooperation in non-traditional security. LSE IDEAS—Dahrendorf forum special report, Apr 2016
27. UNOCHA (2009) Human security in theory and practice application of the human security concept and the United Nations Trust Fund for Human Security. https://www.unocha.org/sites/dms/HSU/Publications%20and%20Products/Human%20Security%20Tools/Human%20Security%20in%20Theory%20and%20Practice%20English.pdf
28. Bell C, Masys AJ (2019) Climate change, extreme weather events and global health security—a lens into vulnerabilities. In: Masys AJ, Izurieta R, Reina M (eds) Global health security: recognizing vulnerabilities, creating opportunities. Springer Publishing
29. Burwell SM, Townsend FF, Bollyky TJ, Patrick SM (2020) Improving pandemic preparedness lessons from COVID-19. Task force report no. 78. https://www.cfr.org/report/pandemic-preparedness-lessons-COVID-19/pdf/TFR_Pandemic_Preparedness.pdf
30. Weick KE, Sutcliffe KM (2007) Managing the unexpected: resilient performance in an age of uncertainty, 2nd edn. Wiley, San Francisco
31. Taleb NN (2007) The black swan: the impact of the highly improbable. Penguin Books Ltd., London
32. Senge P (1990) The fifth discipline: the art and practice of the learning organization. Doubleday Currency, New York
33. Urry J (2002) The global complexities of September 11th. Theory Cult Soc 19(4):57–69
34. Masys AJ (2012) The emergent nature of risk as a product of 'heterogeneous engineering—a relational analysis of the oil and gas industry safety culture'. In: Bennett S (ed) Innovative thinking in risk, crisis and disaster management. Gower Publishing, UK
35. Masys AJ (2018) Complexity and security: new ways of thinking and seeing. In: Masys AJ, Lin L (eds) (2018) Asia/Pacific security challenges—managing black swans and persistent threats. Springer Publishing
36. Goldin I, Mariathasan M (2014) The butterfly defect: how globalization creates systemic risks and what to do about it. Princeton University Press, Princeton
37. Tavanti M, Stachowicz-Stanusch A (2013) Sustainable solutions for human security and anti-corruption: integrating theories and practices. Int J Sustain Hum Secur (IJSHS) 1:1–17. https://repository.usfca.edu/cgi/viewcontent.cgi?article=1018&context=pna
38. Ibn-Mohammed T, Mustapha KB, Godsell J, Adamu Z, Babatunde KA, Akintade DD, Acquaye A, Fujii H, Ndiaye MM, Yamoah FA, Koh SCL (2021) A critical analysis of the impacts of COVID-19 on the global economy and ecosystems and opportunities for circular economy strategies. Resour Conserv Recycl 164:105169
39. GHSI (2019) Global health security index-building collective action and accountability. https://www.ghsindex.org/wp-content/uploads/2020/04/2019-Global-Health-Security-Index.pdf
40. Hiscott J, Alexandridi M, Muscolini M, Tassone E, Palermo E, Soultsioti M, Zevini A (2020) The global impact of the coronavirus pandemic. Cytokine Growth Factor Rev 53(2020):1–9
41. Masys AJ (ed) (2016) Disaster forensics: understanding root cause and complex causality. Springer Publishing

Exploring Sensemaking: A View Through the COVID-19 Pandemic

Anthony J. Masys

Abstract OECD (A systemic resilience approach to dealing with Covid-19 and future shocks, p. 14, 2020 [1]) argues that 'The excuse that dangers are clear only in hindsight does not stand up to objective scrutiny. Major simulation exercises in OECD countries predicted accurately how a crisis like Covid-19 could unfold, but they were not acted on, or not sufficiently, judging by what has happened'. Heyman et al. (Lancet 385:1884–1901, 2015 [2]) argue that, "the world is ill-prepared" to handle any "sustained and threatening public-health emergency". The complex threat and risk landscape that shapes our security and safety experience is characterized by Volatility, Uncertainty, Complexity and Ambiguity (VUCA) conditions. For example the recent COVID 19 pandemic reflects how *unexpected events often audit our resilience* (Weick and Sutcliffe in Managing the unexpected: resilient performance in an age of uncertainty. Wiley, San Francisco, 2007 [3]). Such *surprising events* often reflect an organization's inability to recognize evidence of new vulnerabilities, sense weak signals or the existence of ineffective countermeasures (Woods in Resilience engineering. Ashgate Publishing, Aldershot, 2006 [4]). Public health emergencies stemming from infectious disease outbreaks are creating a serious threat to global health security and have significant National Security implications. Viewed as a black swan or an elephant in the room, this pandemic point to the requirement to be sensitive to the extremes: to those events that lie outside of what we consider predictable (Masys in Homeland security cultures. enhancing values while fostering resilience. Rowman & Littlefield International, London/Lanham, 2018 [5]; Masys in Disaster forensics: understanding root cause and complex causality. Springer Publishing, 2016 [6]; Taleb in The black swan: the impact of the highly improbable. Penguin Books Ltd., London, 2007 [7]). Weick [8] refers to sensemaking in terms of '… how we structure the unknown so as to be able to act in it. Sensemaking involves coming up with a plausible understanding—a map—of a shifting world; testing this map with others through data collection, action, and conversation; and then refining, or abandoning, the map depending on how credible it is' (Ancona in Handbook of teaching leadership. Sage, Thousand Oaks, pp. 3–20, 2011 [9]). Using

A. J. Masys (✉)
College of Public Health, University of South Florida, Tampa, FL, USA

International Centre for Policing and Security, University of South Wales, Cardiff, UK the COVID-19 pandemic as a backdrop and context for this chapter, sensemaking thought leadership will be explored leveraging the seminal work of Weick [8].

Keywords Sensemaking · Critical systems thinking · COVID-19 · National security · Strategic planning

1 Introduction

The complex threat and risk landscape that shapes our security and safety experience is characterized by Volatility, Uncertainty, Complexity and Ambiguity (VUCA) conditions. For example the recent COVID 19 pandemic reflects how *unexpected events often audit our resilience* [3]. Such *surprising events* often reflect an organization's inability to recognize evidence of new vulnerabilities, sense weak signals or the existence of ineffective countermeasures [4]. Such public health emergencies stemming from infectious disease outbreaks are creating a serious threat to global health security and have significant National Security implications. Viewed as a black swan or an elephant in the room, this pandemic point to the requirement to be sensitive to the extremes: to those events that lie outside of what we consider predictable [5–7].

As described in Chapter "The Security Landscape—Systemic Risks Shaping Non-traditional Security", the threat and risk landscape challenges regional and global security along such lines as: national security; energy security; water security; food security; health security; human security; environmental security; economic security. Considering the current COVID-19 pandemic, as of November 2020, the COVID-19 pandemic has seen upwards of 60 million cases, 1.4 million deaths, and has significantly impacted the global economy and affected the most vulnerable. The impacts have seen the closure of borders, economic disruptions and failures, strained and overwhelmed health care systems, failure of supply chains all of which are contributing to a human and national security issue. The COVID-19 revealed the systemic vulnerabilities resident within our global health security system.

Sensing the threat and risk landscape that characterizes the security environment requires an understanding of the inherent systems, interdependencies, nonlinearity across the security domain.

The COVID-19 pandemic has tested and left wanting the global ability to respond to such a threat. The emergence of such disease outbreaks and their influence globally has sparked a renewed attention on global health security and hence on sensemaking for security and safety. Recent outbreaks characterize the "new normal" and has unveiled major deficiencies in preparedness, response and recovery initiatives. OECD [1, p. 14] argue that 'The excuse that dangers are clear only in hindsight does not stand up to objective scrutiny. Major simulation exercises in OECD countries predicted accurately how a crisis like Covid-19 could unfold, but they were not acted on, or not sufficiently, judging by what has happened'.

Things that have never happened before, happen all the time. [10, p. 12]

This quote from Sagan certainly resonates with both natural and man-made disasters as described in Masys [5] and points to systemic risks that lie dormant or latent within our national and global societal systems. Within the context of COVID-19 pandemic the question arises:

> How do we make sense of weak signals that cue us to the emergence of a security or safety vulnerability or disaster?

2 Sensemaking

When we consider the security landscape under VUCA conditions, the words of Woods [4, p. 316] certainly resonate. He asks the question:

> How do people detect that problems are emerging or changing when information is subtle, fragmented, incomplete or distributed across different groups involved in production processes and in safety management. Many studies have shown how decision makers in evolving situations can get stuck in a single problem frame and miss or misinterpret new information that should force re-evaluation and revision of the situation assessment ....

The COVID-19 pandemic has had and continues to have significant impact on our societal systems. We have seen the closure of borders, economic disruptions and failures, strained and overwhelmed health care systems, failure of supply chains all of which are contributing to a human and national security issue. Despite the shock, there were plenty of warning signs. Consider the following recent outbreaks: SARS (2003), H1N1 (2009), MERS (2011), Ebola (2014–2016), all can be viewed as dress rehearsals for COVID-19 pandemic [11]. With this in mind, there is an urgent need to re-examine our exposure to global health security risks and vulnerabilities, in order to transform our understanding and perception of security. This essentially questions how we **make sense** of global security risks, vulnerabilities and impacts.

The complex landscape of sensemaking theories/definitions is captured well in Brown et al. [12]. Brown et al. [12, pp. 266–268] argues that '... there is no single agreed definition of 'sensemaking'. There is, though, an emergent consensus that sensemaking refers generally to those processes by which people seek plausibly to understand ambiguous, equivocal or confusing issues or events ... there is no consensus on whether sensemaking is best regarded primarily as sets of individual-cognitive (e.g. schemata, mental maps), collective-social (interactions between people) or specifically discursive (linguistic/communicative) processes'.

Within the context of crisis (as described in [13]), Fig. 1 depicts some foundational perspectives on sensemaking that capture key features of crisis management and crisis leadership described in this edited volume.

As described by Jones [14]:

> **Weick**'s focus has been organizational activity (collective), and the location of sensemaking is internalized as representation of collective meaning. **Dervin** has a clear individual and hermeneutic approach, on the individual's situation and their internalized subjective experience of it. **Klein**'s focus is the individual mental model (frame) applied to an external context or activity (how external data is represented). **Russell**'s information theoretic view

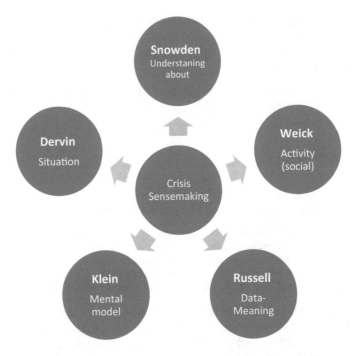

Fig. 1 Crisis sensemaking perspectives (modified and contextualized from [14])

establishes sensemaking as a collective location (an information world) largely in the service of interpreting external data. **Snowden**'s more evolutionary model considers sensemaking a knowledge production activity, using data toward a shared understanding of problem areas (which I call "understanding about" as a unit of analysis).

These perspectives emerge and certainly apply to the shaping of COVID risk communication as described in Sanders [15]. As well, Boin et al. [16] describe information quantity and quality with respect to government decision making pertaining to the COVID-19 pandemic (certainly cross cutting the 4 sensemaking perspectives). In this sense, all these perspectives play a role in sensemaking systemic risks within global and national societal systems in support of crisis management. For a more detailed description of these perspective see Weick [8], Weick et al. [17], Dervin [18], Russell et al. [19], Snowden [20].

Much of the thought leadership associated with sensemaking is rooted in the work of Weick [8]. Weick introduced the term 'sensemaking' to explain how we structure and act under conditions replete with unknowns. In short, 'the making of sense' (Weick [8], p. 4). Weick [8] refers to sensemaking in terms of '… to how we structure the unknown so as to be able to act in it. Sensemaking involves coming up with a plausible understanding—a map—of a shifting world; testing this map with others through data collection, action, and conversation; and then refining, or abandoning, the map depending on how credible it is' [9].

Ancona [9] describes sensemaking as an "… activity that enables us to turn the ongoing complexity of the world into a situation that is comprehended explicitly in words and that serves as a springboard into action" [17, p. 409]. Thus sensemaking involves—and indeed requires—an articulation of the unknown'. It essentially is about making sense of complexity. The failure of sensemaking and its implication on accidents and disasters is well documented [21–25]. Cited in Stern [26, p. 48], Alberts and Hayes [27, p. 102] argue that '… sensemaking is much more than sharing information and identifying patterns. It goes beyond what is happening and what may happen to what can be done about it'. It is about creating an emerging picture of the crisis, developed through data collection, analysis, experience and dialogue across a diverse network stakeholders and participants.

Within the context of crisis and disaster management, as cited in Ntuen et al. [28, p. 4], Leedom [29] notes that sensemaking is a multidimensional process of developing an **operational understanding** and awareness within a complex and evolving task domain.

> **Cognitively**, it can be seen as a process of collecting, filtering, interpreting, framing, and organizing available information into actionable knowledge for decision-making. **Socially**, it can be seen as process of reconciling and integrating multiple stakeholder perspectives into a common operational vision that is driven by a specific goal. **Organizationally** it can be seen as the process of building up appropriate bodies of staff expertise, equipping those bodies with effective information systems and collaboration technology, and efficiently structuring the knowledge management and decision making capabilities of those bodies. [28, p. 4]

This **operational understanding** and awareness is central to crisis management and certainly resonates with the notion of an emergence of a crisis event characterized by a 'disruptive ambiguity' [17, p. 413]. It points to the underlying threats, risks and vulnerability inherent within the security and safety landscape. This operational understanding is reflected in the findings associated with risk communication and government decision making as pose by Sanders [15] and Boin et al. [16].

3 Discussion: Systemic Risks and Sensemaking

Disaster events characterized by Volatility, Uncertainty, Complexity and Ambiguity (VUCA) require leaders and planners to engage in making sense of the inherent dynamics. As articulated in Nowling and Seeger [30], 'they scan their environments for cues of changes and operational threats. After identifying the threat or change, they communicate and work collaboratively to determine the most plausible meaning of these cues [24]. They then enact decisions and actions designed to reduce risk and resolve organizational challenges. When this process is deficient, Weick adds, even the most well-managed organizations can escalate their exposure to risks of crisis'. Sanders [15] describes this well in her analysis.

Under such VUCA conditions, Weick et al. [17, p. 415] describes that 'Sensemaking is not about truth and getting it right. Instead, it is about continued redrafting of an emerging story so that it becomes more comprehensive, incorporates more of the observed data, and is more resilient in the face of criticism'. In this sense, sensemaking is an iterative process of reflective practices. As a social construct, sensemaking is 'an on-going accomplishment that emerges from efforts to create order and make retrospective sense of what occurs' (cited in Weiser [31, p. 3]).

The COVID-19 pandemic provides an apropos backdrop to operationalize the concept of sensemaking. Kalkman [32] described how '… when people are confronted with situations that are ambiguous, chaotic, or simply unexpected, they have to make (renewed) sense of their environment [17, 33, 8]. They try to overcome their surprise and aim to restore their understanding of what is happening around them'. This certainly resonates with the global experience of the COVID-19 pandemic and reifies with crisis communication as discussed in Sanders [15] and Boin et al. [16].

Ala-Laurinaho et al. [34] describes the process aspects of sensemaking. Although rooted in individual mental models [17] and identity construction [35], '… the process of making sense is primarily stated as being an active conversational process with iterations of action and interpretation [17, 36, 37]' [34, p. 368]. Within this 'social' context pertaining to crisis management/crisis leadership defined in Boin et al. [13], Combe and Carrington [38] describes the challenges of sensemaking in crisis conditions. Combe and Carrington [38] argues that '… leaders have to think and problem solve in the context of a novel ambiguous situation involving time pressure and stress while interacting with others in management teams … The key task for leaders in such situations is to develop a mental model, based on their schemas, consisting of causal beliefs for understanding and responding to the crisis [8]'. What this leads to is '… the social construction of sensemaking embraces the collective meaning, individual cognitive processes, temporal dimensions (retrospective and future oriented), the communication of meaning' [38]. This social construction stemming from discrepant cues and retrospective [8] analysis is captured in Maitlis and Sonenshein [39]. It is about the emergence and development of plausible meaning and interpretation stemming from cues. Although framed in terms of a social construction, Fig. 1 does point to the various perspectives in which sensemaking (activity, data-meaning, mental models, situation and understanding about) can be viewed to support crisis management.

4   Crisis, Threats, Risks, Vulnerabilities and Sensemaking

The identification of threats, risks and systemic risks is a significant challenge amongst crisis managers. This is particularly relevant in the face of 'creeping crisis' [40]. Sensemaking is thereby a critical component of crisis management and crisis leadership. As noted in Crayne and Medeiros [41, p. 2] 'Research has suggested that inadequate sensemaking can exacerbate crisis conditions and lead to catastrophic outcomes. For example, akin to the present issue of COVID-19, Weick [8] noted that challenges in sensemaking distributed across multiple institutions were directly related to the Centers for Disease Control's initial misdiagnosis of the West Nile virus during its spread through New York City in the late 1990s. The errors made during this time and an inability to develop comprehensive understanding of the issue resulted, in Weick's [42] view, in the proliferation of a virus that eventually infected and caused harm to millions of Americans. The preponderance of evidence suggests that sensemaking is an essential element to successful navigation of crisis events (see [33] for a review), and that those who take ownership of that process have significant and direct influence over the success of any crisis response'.

Weak signal detection pertaining to risks, threats and vulnerabilities associated with 'creeping crisis [40] is essential to reducing uncertainty and supporting implementation of mitigation and response strategies [30, 43]. As cited in Nowling and Seeger [30] the '... development of a crisis involves a failure to recognize, receive or attend to a threat signal ... The ability to recognize and respond to risk signals is critical for crisis and strategic communicators seeking to maintain or restore an organization's normal function ... Often, these signals are weak, difficult to notice, and hard to interpret'. Facilitating the qualities of crisis leadership (see this edited volume), sensemaking becomes a central and critical component where it is characterized as an ongoing emergent capability and capacity. This has been recognized in countless disaster and crisis events such as Deepwater Horizon, Hurricane Katrina, the Ebola outbreak (2014–2016).

Germane to the context of COVID-19 pandemic, Rubin and de Vries [44] describes how the sensemaking analytical approach associated with Cynefin offers four decision-making frames or domains in which to help leaderships to perceive situations characterized by VUCA. Such a systems thinking and complexity lens to pursue learning and solution navigation is described in detail in Angeli and Montefusco [45] and Sturmberg and Martin [46].

Understanding the threat and risk landscape is facilitated through a process of continued learning and engaging of the crisis situation resulting in the emerging plausible story. Sensemaking is the engaging framework in which facilitates the management of crisis. Failure of effective sensemaking and decision making exacerbates a crisis across the disaster management spectrum (mitigation, preparedness, response, recovery). Examples across the COVID-19 experience have been cited.

5 Conclusion

As noted in Christianson and Barton [47] 'when people encounter surprising or confusing events, they engage in sensemaking to answer the questions, 'what's the story?' and 'now what?' [17]. Sensemaking is a socially constructed process in which individuals interact with their environment and with others to create meaning and enable action'. The COVID-19 pandemic represents a global health security, national security and human security disaster. The characteristics that are reflected in the COVID-19 epidemic are Volatility, Uncertainty, Complexity and Ambiguity (VUCA). As argued by Weick [48] 'the sensemaking perspective is particularly relevant in crisis situations characterized by ambiguity, confusion, and feelings of disorientation' (cited in Maitlis and Sonenshein [39]).

In this chapter we have examined the characteristics of sensemaking through the contextual lens of the COVID-19 pandemic. It is recognized that although different perspectives of sensemaking are described in the literature, within the context of crisis management it points to the questions 'what is the story?' and 'now what?'. This is all about crisis sensemaking.

References

1. OECD (2020) A systemic resilience approach to dealing with Covid-19 and future shocks. In: New approaches to economic challenges (NAEC), 28 Apr 2020
2. Heyman et al (2015) Global health security: the wider lessons from the west African Ebola virus disease epidemic. Lancet 385(9980):1884–1901
3. Weick KE, Sutcliffe KM (2007) Managing the unexpected: resilient performance in an age of uncertainty, 2nd edn. Wiley, San Francisco
4. Woods DD (2006) Essential characteristics of resilience. In: Hollnagel E, Woods DD, Leveson N (eds) Resilience engineering. Ashgate Publishing, Aldershot, pp 21–34
5. Masys AJ (2018) Designing high reliability security organizations for the homeland security enterprise. In: Siedschlag A, Jerkovic A (eds) Homeland security cultures: enhancing values while fostering resilience. Rowman & Littlefield International, London/Lanham
6. Masys AJ (ed) (2016) Disaster forensics: understanding root cause and complex causality. Springer Publishing
7. Taleb NN (2007) The black swan: the impact of the highly improbable. Penguin Books Ltd., London
8. Weick KE (1995) Sensemaking in organizations. Sage, CA, Thousand Oaks
9. Ancona DL (2011) Sensemaking: framing and acting in the unknown. In: Handbook of teaching leadership. Sage, Thousand Oaks, pp 3–20
10. Sagan S (1993) The limits of safety: organizations, accidents, and nuclear weapons. Princeton University Press, Princeton
11. Hiscott J, Alexandridi M, Muscolini M, Tassone E, Palermo E, Soultsioti M, Zevini A (2020) The global impact of the coronavirus pandemic. Cytokine Growth Factor Rev 53(2020):1–9
12. Brown AD, Colville I, Pye A (2015) Making sense of sensemaking in organization studies. Organd Stud 36(2):265–277
13. Boin A, 't Hart P, Stern E, Sundelius B (2005) The politics of crisis management public leadership under pressure. Cambridge University Press
14. Jones PH (2015) Sensemaking methodology: a liberation theory of communicative agency. https://www.epicpeople.org/sensemaking-methodology/
15. Sanders KB (2020) British government communication during the 2020 COVID-19 pandemic: learning from high reliability organizations. Church Commun Cult 5(3):356–377
16. Boin A, Lodge M, Luesink M (2020) Learning from the COVID-19 crisis: an initial analysis of national responses. Policy Des Pract 3(3):189–204
17. Weick KE, Sutcliffe KM, Obstfeld D (2005) Organizing and the process of sensemaking and organizing. Organ Sci 16(4):409–421
18. Dervin B (1999) Chaos, order and sense-making: a proposed theory for information design. Inf Des 35–57
19. Russell DM, Stefik MJ, Pirolli P, Card SK (1993) The cost structure of sensemaking. Paper presented at the INTERCHI'93 conference on human factors in computing systems, Amsterdam, Apr 1993

20. Snowden DJ (2005) Multi-ontology sense making: a new simplicity in decision making. Inform Primary Care 13(1):45–54
21. Macrae C (2007) Interrogating the unknown: risk analysis and sensemaking in airline safety oversight. LSE discussion paper no: 43, May 2007. https://eprints.lse.ac.uk/36123/1/Disspaper43.pdf
22. Snook SA (2000) Friendly fire: the accidental shootdown of US black hawks over Northern Iraq. Princeton University Press, Oxford
23. Turner B (1978) Man-made disasters. Wykeham, London
24. Weick KE (1993) The collapse of sensemaking in organizations: the Mann Gulch disaster. Adm Soc 38(4):628–652
25. Masys AJ (2010) Fratricide in air operations: opening the black box-revealing the social. PhD Dissertation, June 2010, University of Leicester, UK
26. Stern EK (2017) Crisis, leadership and extreme contexts. In: Holenweger M, Jager MK, Kernic F (eds) Leadership in extreme situations. Springer Publishing
27. Alberts DS, Hayes RE (2003) Power to the edge. Command and Control Research Program, Washington
28. Ntuen CA, Munya P, Trevino M (2006) An approach to collaborative sensemaking process topic: social domain issues, C2 concepts and organizations, C2 analysis. In: 11th ICCRTS coalition command and control in the networked era
29. Leedom DK (2004) The analytic representation of sensemaking and knowledge management with a military C2 organization. Final Report AFRL-HE-WPTR2004-0083. Human effectiveness Directorate, WPAFB, OH
30. Nowling WD, Seeger MW (2020) Sensemaking and crisis revisited: the failure of sensemaking during the Flint water crisis. J Appl Commun Res 48(2):270–289
31. Weiser A-K (2020) The role of substantive actions in sensemaking during strategic change. J Manag Stud
32. Kalkman JP (2020) Sensemaking in crisis situations: drawing insights from epic war novels. Eur Manag J 38:698–707
33. Maitlis S, Christianson M (2014) Sensemaking in organizations: taking stock and moving forward. Acad Manag Ann 8:57–125
34. Ala-Laurinaho A, Kurki A-L, Abildgaard JS (2017) Supporting sensemaking to promote a systemic view of organizational change—contributions from activity theory. J Change Manag 17(4):367–387
35. Thurlow A, Mills JH (2015) Telling tales out of school: sensemaking and narratives of legitimacy in an organizational change process. Scand J Manag 31(2):246–254
36. Balogun J, Johnson G (2005) From intended strategies to unintended outcomes: the impact of change recipient sensemaking. Organ Stud 26(11):1573–1601
37. Lawrence P (2015) Leading change—insights into how leaders actually approach the challenge of complexity. J Chang Manag 15(3):231–252
38. Combe IA, Carrington DJ (2015) Leaders' sensemaking under crises: emerging cognitive consensus over time within management teams. Leadersh Q 26(2015):307–322
39. Maitlis S, Sonenshein S (2010) Sensemaking in crisis and change: inspiration and insights from Weick (1988). J Manage Stud 47:3
40. Boin A, Ekengren M, Rhinard M (2020) Hiding in plain sight: conceptualizing the creeping crisis. Risk Hazards Crisis Public Policy 11(2)
41. Crayne MP, Medeiros KE (2020) Making sense of crisis: charismatic, ideological, and pragmatic leadership in response to COVID-19. Am Psychol
42. Weick KE (Ed) (2005) Managing the unexpected: complexity as distributed sensemaking. In: Uncertainty and surprise in complex systems. Springer, Berlin, Germany, pp 51–65
43. Slovic P (2000) The perception of risk. Everscan, London, pp 220–231
44. de Rubin O, Vries DH (2020) Diverging sensemaking frames during the initial phases of the COVID-19 outbreak in Denmark. Policy Des Pract 3(3):277–296
45. Angeli F, Montefusco A (2020) Sensemaking and learning during the Covid-19 pandemic: a complex adaptive systems perspective on policy decision-making. World Dev 136:105106

46. Sturmberg JP, Martin CM (2020) COVID-19—how a pandemic reveals that everything is connected to everything else. J Eval Clin Pract 26:1361–1367
47. Christianson MK, Barton MA (2020) Sensemaking in the time of COVID-19. J Manag Stud
48. Weick KE (1988) Enacted sensemaking in crisis situations. J Manag Stud 25:305–317

Sensemaking Under Conditions of Extreme Uncertainty: From Observation to Action

Christian Fjäder

Abstract The contemporary strategic operating environment is increasingly characterized by increasing complexity and dynamic change, leading into new vulnerabilities and uncertainty. Whilst global information and transparency have reached unparalleled levels, we still seem to be taken by surprise with sudden shocks and crisis. In some cases this is due to the deep or extreme uncertainty, i.e. in case of events that are so unique that they are genuinely unprecedented or extremely rare (think of a 'planet killer' meteorite striking Earth), or high impact—low probability risks that do not get prioritized by political decision-makers as preparedness priorities. The COVID-19 pandemic appears to be a good example of the latter. Whilst experts were warning governments (for instance the Global Preparedness Monitoring Board in a September 2019 report [1]) about an influenza pandemic involving a high-impact respiratory pathogen and criticized them about the lack of preparedness for such an event, the exponential spread of COVID-19 took governments around the world by surprise. The initial excuse was that COVID-19 was an unexpected 'black swan', which it of course was not. Either the thinking as it relates to pandemic threats is in too linear terms, or the message from the experts was not sufficiently convincing to political decision-makers to spring into action. This begs the question—why do we either ignore such risks or fail to take decisive action in face of a large-scale disaster? The purpose of this chapter is to review and examine methods and approaches that could enable governments and organizations to integrate security and preparedness decision-making from monitoring and detection to action through improved sensemaking. In other words, it seeks a path to being less taken surprise by the perils of deep uncertainty and taking the appropriate actions in a timely and sufficient manner.

Keywords Sensemaking · Uncertainty · COVID-19 · Black swan

C. Fjäder (✉)
Geostrategic Intelligence Group (GIG), Esbo, Finland
e-mail: christian.fjader@geostrat.fi

The Finnish Institute of International Affairs (FIIA), Helsinki, Finland

1   Introduction

The contemporary strategic operating environment is increasingly characterized by increasing complexity and dynamic change, leading into new vulnerabilities and uncertainty. Whilst global information and transparency have reached unparalleled levels, we still seem to be taken by surprise with sudden shocks and crisis. In some cases this is due to the deep or extreme uncertainty, i.e. in case of events that are so unique that they are genuinely unprecedented or extremely rare (think of a 'planet killer' meteorite striking Earth), or high impact—low probability risks that do not get prioritized by political decision-makers as preparedness priorities. The COVID-19 pandemic appears to be a good example of the latter. Whilst experts were warning governments (for instance the Global Preparedness Monitoring Board in a September 2019 report) about an influenza pandemic involving a high-impact respiratory pathogen and criticized them about the lack of preparedness for such an event, the exponential spread of COVID-19 took governments around the world by surprise. The initial excuse was that COVID-19 was an unexpected 'black swan', which it of course was not. Either the thinking as it relates to pandemic threats is in too linear terms, or the message from the experts was not sufficiently convincing to political decision-makers to spring into action. This begs the question—why do we either ignore such risks or fail to take decisive action in face of a large-scale disaster? Yet, the sheer speed and scale of events is such that decisions need to be made fast, including tough ones. Hence, the chain from detection to action cannot be too long, which challenges the hierarchical escalation procedures and consensus decision-making. Secondly, the probability-based risk management approach clearly is defunct, or at least inadequate, as an approach to events in the deep uncertainty category. Also, whilst strategic foresight and horizon scanning activities are still very useful in many ways, they must facilitate taking action better. Strategic intelligence analysis techniques could offer a partial solution for gearing information into a framework that is understandable to decision-makers and suggests concrete pathways for action. Finally, due to the dynamic pace of change and exponential impact of sudden shocks, organizations and governments should develop improved capabilities for continuous situational awareness that is sufficiently integrated into decision-making processes to enable rapid responses to arising threats.

The purpose of this chapter is to review and examine methods and approaches that could enable governments and organizations to integrate security and preparedness decision-making from monitoring and detection to action through improved sense-making. In other words, it seeks a path to being less taken surprise by the perils of deep uncertainty and taking the appropriate actions in a timely and sufficient manner.

2 Extreme Uncertainty: Survival in a World of Surprises

The world we are living in is arguably becoming more complex, dynamic and uncertain, which means that it is increasingly difficult to anticipate significant changes in the strategic environment in order to prepare against their negative impact in time. As a consequence, it seems we are increasingly taken by surprise of events that we in principal should have been able to anticipate in advance, even when such events seem evident in hindsight. The result is what could be called "extreme uncertainty", the significantly reduced ability to ascertain between the unlikely and implausible. The global COVID-19 crisis is merely one example of such events.

This problem of extreme uncertainty was also made famous by Nassim Nicholas Taleb, the author of the best-selling book—"The Black Swans", who used the metaphor of black swans to the propensity of trying to forecast hard to predict extreme events, i.e. low probability—extreme impact events, or assigning probabilities to rare events using scientific methods. Attempts to utilise scientific methods to measure such extreme outliers thus, creates false hope of prediction on events that are genuinely explainable only in the hindsight [2]. Displaying a similar spirit, the U.S. Secretary of Defence Donald Rumsfeld famously stated in February 2002 in a Defence Department's media briefing, when asked about evidence of weapons of mass destruction in Iraq, that "There are known knowns. There are things we know that we know. There are known unknowns. That is to say, there are things that we now know we don't know. But there are also unknown unknowns. There are things we do not know we don't know". Whilst initially ridiculed about this seemly cryptic statement, Rumsfeld has also received some credit for drawing attention to an important challenge, suggesting that we might be unprepared to completely unexpected catastrophic events that are beyond parameters that would be considered 'normal' simply because we are not aware of even their existence or possibility to occur. The same line of thought is behind the concepts of TUNA (turbulent, uncertain, novel, ambiguous), VUCA (volatility, uncertainty, complexity, ambiguity) and BANI (brittle, anxious, non-linear, and incomprehensible).

This section will investigate the fundamental drivers for this arguably increasing extreme uncertainty and explore the fundamentals for strategies to reduce the possibilities of being surprised about potentially catastrophic events in the operating environment.

2.1 The Increasing Complexity and Dynamism—The Global Value Chains

Complexity in the contemporary world extends well beyond the interdependencies of the national economies and the complex networks of global economic transactions, but they are perhaps the most demonstrative example of the dynamics at play behind the emerging uncertainties and systemic risk of the contemporary world. The sheer scale and volume of global economic transactions and other interaction that criss-crosses the globe with little regard to national borders provides immense opportunities to businesses of all size and kind. The global value and supply chains which fuel the global economy are the extreme manifestation of the principle comparative advantage, making it possible for a business of any size to sell its product and services globally, as long as it has the competitive advantage over others. These global arteries channel immense flows of people, goods, finance, data and ideas that impact human life in diverse ways from the economy to culture and education. However, they also bring with them a level of complexity and dynamism that creates new vulnerabilities and risks as disruptive effects can emerge from great distances and from locations that would at first thought reward little importance and following events that would seem equally unimportant. Now a flood in Thailand, typhoon in the Philippines, an act of piracy off the Somalian coast or a cyber-attack in India can cause almost immediate and devastating impact in the other side of the world. The inherent complexity of such complex adaptive systems is a subject of complexity science, for instance the "chaos theory", which introduced the concept of "butterfly effect", depicting the effects of butterfly flapping its wings somewhere causing a hurricane or typhoon in the other side of the world [3, p. 201]. Globalisation scholars Ian Goldin and Mike Mariathasan in their book—*The Butterfly Defect: How Globalization creates systemic risks, and what to do about it* [4]—point out that globalization is a double edged sword in a sense that in addition to creating a lot of good, it also creates new systemic risks though unprecedented global connectedness that can cause cascading effects from seemingly unrelated micro-systems to other interdependent and dependent systems, thus causing system-level impacts that overwhelm systemic resilience. This type of "butterfly defect" is a good example of the increasing complexity and dynamic change that causes extreme uncertainty and make us prone for surprises.

In addition to the volatility of the global system, it naturally carries a "dark side" with it. Hence, in addition to the positive effects of globalisation, for instance the global flows of trade and capital that have accelerated the growth of the global economy, it has also facilitated a broad set of illicit flows operated by international criminal and terrorist organisations to distribute drugs, arms and other illicit goods globally. Consequently, globalisation has also changed the concept of security, shifting its emphasis from traditional protective security and law enforcement towards flow security—securing the critical flows carrying critical goods, services and information and the nodes that provide access to them (see for instance: [5–7]). This new dynamic brings with it a new reality of uncertainty.

As indicated earlier, the turbulence and volatility of the global operating environment has also drawn attention in executive education programmes and business literature. The Oxford University Executive Education programme refers to this condition as the "Turbulent—Uncertain—Novel—Ambiguous (TUNA)", whilst in business literature the widely used term is VUCA—volatile, uncertain, complex, ambiguous. Regardless of the differences in vocabulary, the intention is the same: the strategic environment in which in particular global organisations operate is subject to constant, yet unpredictable and sudden changes with impacts that are hard to predict or even understand. Hence, it is increasingly difficult to predict what exogenous events might occur and how they might impact the organisation and its objectives. This is what is essentially called "extreme" or "deep" uncertainty. In order to enable prepared-ness and resilience against such extreme uncertainty, a deeper and more holistic understanding of the relationship and difference between the concepts of "risk" and "uncertainty" is required.

2.2 The Difference Between "Risk" and "Uncertainty"

The term "risk" is rather vaguely utilised in colloquial context, generally referring to the possibility of an event considered as undesirable may (or may not) occur, for instance a traffic accident, sports injury or bankruptcy. Moreover, there is no universal scientific definition for "risk" [3, p. 38]. The term" uncertainty" is, in a similar manner, often rather vaguely defined in colloquial context. Most frequently "uncertainty" simply either refers to the limits of our knowledge in regard to the possibility of sudden and unexpected (undesired) events—such as the so called 'black swans' or 'unknown unknowns' (see more below)—or as an alternative—a sub-category of risk. The term is also commonly used in the context of uncertainty of measurement in many fields, including statistical approaches to risk management (stemming from scientific and statistical approaches, see more below) In relation to risk management, "uncertainty" has been referred to events with "unknown outcomes with unknow probability law" [3, p. 39].

However, whilst the concepts of risk and uncertainty are related, they are not the same. In order to distinguish between the two and the significance of the differences, it is important to have a basic understanding of the theories of *risk* and *uncertainty*. First of all, "risk" carries both, very specific and very fluid meanings. For instance, the Bayes' theorem states that agents utilize probabilistic assessments about the likelihood of events based on observations on their past occurrence to understand their operational environment. The same underlining concept is reflected in international risk management standards. The ISO 31000 (2018), for instance, defines risk as "effect of uncertainty on objective", clarifying that "effect" should be understood as "a deviation from the expected", whether this is positive, negative or both, in any level of objective relevant in the context (ISO 3001). This specific and systematic view of "risk" is the basis of modern risk management in organisations around the world, across the industries and in all kinds of organisations, regardless whether they are private or public sector entities, for profit or not for profit and any and all other organisations following international risk management standards. Risk Management as a discipline[1] essentially aims to mitigate or manage the impact of such events by measuring the probability and impact of known risks, based on our knowledge on the occurrence of similar events, and then identify measures to mitigate (reduce), manage (limit) or transfer (principally to insurance) the risk. Ultimately this standard

---

[1]ISO 3001 [8] defines "risk management" as "coordinated activities to direct and control an organisation with regard to risk".

risk management approach aims to manage the organisation's exposure to (harmful) risk by limiting it to an acceptable level—called "residual risk". The general idea is to reduce the organisation's risk exposure to an optimal level. Exogenous risk, however, can never be eliminated entirely and hence, becoming "risk free" is not possible. This creates a permanent space for "uncertainty". The relationship between "risk" an "uncertainty" is, however, somewhat more complex.

Knight [9], one of the prolific theorists of risk, distinguished the differences between "risk" and "uncertainty" in his seminal book *Risk, Uncertainty and Profit*, by arguing that whilst "risk" is observable and measurable, "uncertainty" operates in the limits of our knowledge, making assigning probabilities impossible.[2] The term "risk" can thus be said to refer to situations where probability and impact of an undesirable event can be determined, because the possible outcomes can be identified, and the past frequency of their occurrence can be determined through observations of past events. Hence, following Knight's thesis, "uncertainty", suggests that possible outcomes are not known to us, or that decision-makers do not hold adequate knowledge or experience concerning the situation, in order to assign probabilities for the possible outcomes or to understand their possible impacts. This in turn leads to the inability to determine the appropriate response based on rational calculations, as well as making decisions following the standard risk management process to either accept, mitigate or transfer the risk. This challenge is multiplicated in a contemporary operating environment characterised by ever increasing levels of interdependence, complexity and dynamism. As a consequence, events or incidents that would seem minor in a narrower concept can unexpectedly create sudden and massive changes, as Goldin and Mariathasan have pointed out in "the Butterfly Defect".

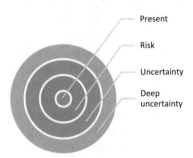

The risk-uncertainty spectrum. *Source* Author

The concept of uncertainty in fact exists in more areas of our life than we can imagine at first thought, including but not excluding in decision-making activities such as, the judicial system (guilty beyond reasonable doubt), medicine (do not resuscitate), as well as odds we take with everyday life decisions (getting married, buying a house, choosing a careers) and, especially in business decisions (where risk is always present and often also represents an opportunity).

---

[2]See Knight [9] and Jarvis [10].

In scientific terms this relates to not only uncertainties in measurement in order to validate the hypotheses, as statistical approaches would suggest, but also to the *ontological* and *epistemological* differences, in other words in relation to what we consider as reality and as evidence of that reality. For the sake of simplicity, one could summarise the differences by stating that the short-term unknow future could be characterised as "risk", whereas in the long-term we are really dealing with "uncertainty". Following this logic, the longer the horizon, the more the balance tilts towards "uncertainty" [3, p. 39].

Present-future view to mitigation. *Source* Author

3 Avoiding Surprise: Establishing Situational Awareness

As has been established thus far, we still have considerable limitations towards detecting and understanding extreme uncertainty. Hence, we seem to be taken by surprise virtually every time a major crisis erupts. Perhaps a more serious short-coming, however, is that even when we do detect a crisis we still fail to act appropriately. This is particularly true in regard to "creeping crises", which escalate gradually, and of which the ongoing COVID-19 pandemic seems be a perfect example of.

This is partially due to our shortcomings in detecting the relevant changes in our operating environment, making sense of them and taking the appropriate actions in reaction. It also stems for the shortcomings of the probabilistic risk management, which is not applicable to deep uncertainty. This section addresses the first part of the puzzle; how can an organisation ensure it senses and identifies relevant changes in its operating environment?

3.1 Sensing and Identifying Relevant Changes in the Operating Environment

There are always infinite number of events and 'signals' in a global operating environment that could potentially have significance to any organisation. Moreover, there are numerous events on daily basis that are clearly negative and even threatening, i.e.

terrorist attacks, kidnapping, natural disasters, riots and mayhem and geopolitical tensions. There may be even small incidents that might not seem consequential at the time, but nonetheless trigger the "butterfly defect" shortly after. One demonstrative example of such events is the young Tunisian fruit vendor Mohamed Bouazizi, whose death triggered the anti-government protests that became a catalyst for the Arab Spring in 2011. Albeit tragic and soon broadcasted on Facebook for everybody to see, the self-immolation of a simple fruit vendor protesting harassment by officials in front of a government building led not only to the crumbling of the regime in Tunisia but the mayhem spread like a wildfire across the region with amazing speed and momentum. The problem with these types of incidents is that unfortunately they become consequential only in the perspective of hindsight, if even then. Are we thus doomed to be surprised by such events for eternity? Whilst we probably never gain the ability to comprehend the significance of such events in real time, we can improve our preparedness against them by detecting them and taking note of them as early as possible, as well as by improving our ability to contextualise such events in the context of other events and signals.

The first step is that the organisation establishes the capability to monitor the strategic environment for events or phenomena that appear novel and thus, such it has no or little experience of (and hence, probably no processes to protect against them). If the phenomenon is previously unknown to the organisation but relevant to its operations, it indicates that it is probably something we should aim to understand for relevance to our organisation. However, before rushing into conclusions, once you have identified a novel phenomenon evaluate it for ambiguity, this confirms that the phenomenon exists independent of ones already known to us. The Tunisian fruit vendor and the Arab Spring case is again a good example for this. Having confirmed that the phenomenon we have detected is novel and exists independent of those we are already aware of; we should evaluate it for the level of complexity. This tells us how much effort we need to dedicate towards understanding it in a level that enables determining the course of action and taking action in reaction to the phenomena. If the phenomenon is relatively simple or understandable, its causes and effects are also relatively easy to identify and evaluate. In such case, a reasonable level of certainty about its implications to the organisation is possible to establish. If the phenomena has a high level of complexity, things get ... well, more complex. This is because complexity itself is complex. Some complex phenomena may have a multitude of causal and correlational interactions between constituent parts, but they do not change especially dynamically. Think of for example an immensely complex technical system such as a space rocket, which obviously is very complex technical design with immense number of different types if interactions required for the desired end result, but at the same time a system designed for predictable reactions with high levels of certainty. Other complex phenomena may have less causal and correlational interactions, but they are constantly subject to change. Just think of different manifestations of human interaction from everyday economic transactions, such as buying groceries, to decisions involved in buying a car or a house, not to mention dating and getting married. Even though people face the equivalent or same choices in the same circumstances more than once, whether it is daily of only twice in life, people nonetheless frequently make very different choices each time the situation is repeated. In brief, humans are adaptive by basic nature and constantly readjusting to the new options opening in front of them. Hence, whilst there are massive amounts of data on how humans react to a certain choice presented to them under certain circumstances, we still have major challenges in predicting their decisions. Whilst there has been significant improvement in such capability, for instance in predicting shopping behaviour, thanks to algorithms, yet the fact that people may be perfectly happy making an irrational choice sets severe limitations to predicting human behaviour. Hence, some phenomena are relatively complex but stable, meaning that they are not especially volatile to (significant) change and can thus, be addressed by systemic analysis following causal and correlational chains step by step. This may be difficult and time-consuming work, but its results are relatively certain. On the other hand, phenomena that are unstable and constantly under change, also in relation to the causal and correlational relations, are more difficult to analyse with reasonable level of certainty. This could easily lead us to conclude that very complex and stable are easier to understand than less complex and more volatile phenomena. Unfortunately, it is not this straightforward. Again, complex and stable take time and effort but the results are less uncertain, whereas complex and volatile involve so many alternative causes and effects that expand to different combinations exponentially so that our analysis is subject to more uncertainty. In sum, an organisation should seek capabilities to **monitor** its operating environment and **identify** events and weak signals that are relevant. In addition, it should be able to **confirm** the **existence and novelty** of such events and **assess** their **complexity** and **volatility.** In order to "keep its eye on the ball", it then needs to set up a capability to **monitor** those it has designated as relevant, but also continue monitoring the operating environment for new events and signals. In sum, it needs to establish a **continuous capability to monitor and analyse** its operating environment.

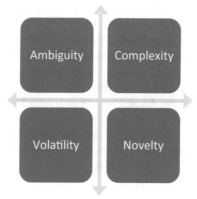

The deadly quadrant of ambiguity, volatility, complexity and novelty. Source *Author*

Whilst there are very few, if any, best practices available for consultation on how to exactly establish this capability, one potential starting point is the *OODA Loop* approach.

3.2 The OODA Loop Approach

One option in such situations is to adapt the OODA loop approach. OODA loop approach was created by United States Air Force Colonel John Boyd for the purposes of improving the responsiveness of military combat operations processes. The OODA loop offered a particular purpose and advantage to fighter pilots who required the ability to quickly assess the situation and decide the appropriate course of action to respond to it. Moreover, in quickly paced air operations this capability had to be constant and hence, required a repetitive process that could be instinctively repeated once the operator had learned it. OODA loop has since been adapted also in law enforcement, rescue services, business and elsewhere where quick decisions are a must for survival as a method for dealing with uncertainty in decision-making under dynamic conditions. OODA stands for Observe, Orient, Decide and Act phases, which are all executable in a very short span. The basic idea is to provide a dynamic and recurring loop that restarts every time that the conditions change. In order to use it effectively, its execution requires consistency: all phases must be repeated until the situation has been solved. It is, however, also intended as being dynamic, i.e. there is a logic for the repetition, which dictates a need to repeat the loop after seeing how actions have taken effect or circumstances have changed, or how the opponent has adapted.

In its military application this meant dynamically countering the opponents act in a manner to maintain a competitive advantage. This is an important point for those planning to use it in an organisation, possibly as a team effort, in circumstances where there is no opponent per se, e.g. in a crisis where there is no human opponent, such as a natural disaster or a pandemic. The competitive and adversarial approach is natural in combative situations where the actor that can perform the loop faster than others will gain a competitive advantage and will hence, probably win the fight. This logic does not necessarily hold in all circumstances and decision-making systems and cultures. For instance, large corporations and other large civilian organisations tend to be more prone toward collaborative and consensus-seeking in their approach. It is, however, a powerful method in sudden and unexpected circumstances where there is a critical need to take action, whether as a matter of mere survival or in order to gain an advantage against competitors facing the same circumstances. As such, it rewards a closer look.

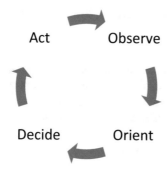

The OODA loop. *Source* Author's adaptation

The *observe* phase focuses on observing and taking in to account the necessary new information about the changing operating environment as a measure of establishing situational awareness. Its purpose is simply to identify and account for all the information that may be relevant.

The *orient* phase, in the simples of possible terms means just that: orienting oneself in the rapidly moving environment. For an individual, such as a fighter pilot, it might mean simply being aware of one's position and the available vectors for changing positions. Thus, it may involve questions, such as; "where am I?", "where can I go safely from here?", "where am I in relation to my objective, my opponent or my team members?". To an organisation or a team, it means having an understanding of what is the organisation's position in the new environment, how it differentiates from its earlier position, where it can lead and determining the parameters for conducting oneself in the new environment.

The *decide* phase is also rather obvious. It involves utilising the information gathered in the earlier two phases in order to form hypotheses about the possible course of action and choosing the option that seems most advantageous towards reaching the desired result.

The *Act* phase completes the circle, whilst restarting the loop again, as suggested earlier. It involves putting into the effect the actions decided in the previous phase and creating some desired effect in the operating environment. Once the action has been carried out and is thought of having had effect, the loop restarts with the *observe* phase in order to reassess the new situation stemming from actions and reactions/effects to them.

In sum, the OODA Loop approach can have utility in a sudden and unexpected adverse situations or events that rewards a rapid and decisive response. It is most effective in uses of first response in fast paced situations, but if the organisation faces a crisis—by default an ongoing event or situation that threatens its survival also in the long-term, one should consider opting for the standard Crisis Management cycle. If the crisis is prolonged or likely to have long-term strategic impact on the organisation or its mission even after the events have terminated, one should consider utilising the resilience approach.

4 Options for Reducing Uncertainty Related to Future Events

Whilst being aware of the relevant changes in the operating environment and having the capability to act on them timely and appropriately is key to survival in the dynamically changing operating environment, being sensitive to long-term change requires improved capabilities for understanding of the future and enables building resilience against deep uncertainty. Strategic foresight is a capability in high demand in a complex, dynamic and uncertain strategic environment.

4.1 Strategic Foresight as Tool for Reducing Uncertainty

Given the inherent uncertainty associated with the future and the complex and dynamic change in a TUNA/VUCA world, strategic foresight approaches have gained in popularity. Whilst predicting long-term futures is implausible, organisations of different kinds have embarked on journeys to increase their understanding of the potential futures in order to be prepared for change proactively. Many organisations have started recognising the value of this capability, rather than simply reacting to change as and when it happens, or worse—after it has already materialised.

Strategic foresight has become increasingly popular across a diverse range of disciplines, policy contexts and strategic decision-making in business to examine the significant uncertainties relating to long-range planning, which is typically anything ranging from 1 to 10 years in time horizon.[3] The most popular approach is *scenario planning*, which has proven useful as a tool to reduce the uncertainty related to the future. Scenario planning emerged following a boom in long-range planning in the 1950s and 1960s by RAND, NASA and others in response to greater need for the ability to forecast major uncertainties related to technology and missions, such as the Apollo program. The long-range planning approach has more recently been dominated by the scenario planning approach, in particular in business. One of the first, and perhaps still best known, is Shell International, which in the 1970s wanted to improve its capabilities in understanding the uncertainties related to the oil crisis and the future of fossil fuels. Scenario planning gained popularity also in government and public sector organisations, especially in the 2000s, as a policy planning tool (e.g. the Singapore Government and governments in Europe).

The contemporary approaches to scenario planning are divergent, depending on the context in which and goals towards which they are produced. Whilst some "best practices" have emerged, the experts point out that there is no one "best" or "right" (and hence, "wrong") method for conducting scenario planning. Instead it is up to the organisation to determine its own approach, as long as it offers a fit for the intended purpose and can be carried out within the organisation's capabilities [11, p. 19].

---

[3] Albeit there are projects that go well beyond this, working on 20, 50 or 100 year time horizons (and even longer).

In general, scenario approaches utilise structured techniques to collect inputs from experts and other participants about "change drivers" in order to produce collective views that can be formulated into multiple divergent views of the future [3, p. 85]. In the Oxford Scenario Planning Approach (OSPA), the process emphasises "reframing" instead of prediction and a strategic learning process that produces alternative "frames" of the future. These "frames" can then be used to "reveal, test and redefine" the "official" or conventional views of the future in the organisation [11, p. 3]. It also emphasises specifically the role of the "scenario learners" and their capabilities to "examine the contextual environment", thus utilising a methodology "... that uses the inherent human capacity for imagining futures to better understand the present situation and to identify possibilities for new strategy" [11, p. 1]. As such, the scenario approach has been sometimes referred to as a "divergent future-gazing method" and a "method of formal storytelling", meaning that "stories" are constructed around the different development paths (negative, positive, neutral) of the "change drivers". These stories are then formed into divergent and plausible scenarios that present different combinations of these drivers as alternative futures [3, p. 85].

One could of course ask why we should be so concerned about the future, especially if we cannot predict it? The power of scenario planning is often said to lay in the path, not the end result. Scenario planning is a powerful method of helping individuals and organisations "to engage with deep uncertainty" [11, p. 64]. Moreover, the purpose is not to be right, but to be ready to face the future. Hence, scenarios may be very useful towards fostering adaptive planning and response [3, p. 85]. Thus, they can also enable the organisation to maintain a positive trajectory in an uncertain future [3, p. 88].

More recently, the scenario approach has gained popularity also in security, ranging from international relations to anti-terrorism and disaster risk and resilience. In this context it has sometimes been referred to as a tool to complement risk management [12].

However, scenarios arguably especially useful in looking beyond risk as a tool for reducing uncertainty. Whereas probabilistic methods of risk management normal statistical distribution on data collected in the past and assume that the causal logic remains unchanged in the future, the value of scenario planning is probably at its highest in the context of surviving and thriving in an increasingly interconnected and turbulent world where connectivity is the key for value, as well as vulnerability and thus, sudden disruptive changes are increasingly common [11, pp. 1, 2]. Consequently, organisations in the various security related disciplines have started paying more attention to emerging and systemic risk instead [11, p. 63].

One demonstrative example is the National Emergency Supply Agency (NESA) of Finland, which has conducted scenario planning projects in 2012 and 2017 in order to enable (long-term) strategic planning of national *Security of Supply* and *Critical Infrastructure Protection (CIP)*. The 2012 edition focused on the time horizon up to 2025. The 2017 edition—*Security of supply: scenarios 2030*—introduced five divergent scenarios, completed as a result of a collaborative effort between national security and foresight experts and the private sector;

1. Global Interdependency
2. Armed Power Politics
3. Blocification and Hybrid Influence
4. Technological World Order
5. The Dominance of the East.

The report clearly states the motives, objectives and limitations of the foresight effort. The primary motive for the effort was stated as the increasing uncertainty in the operating environment (both global and in the near region) and its impact on security of supply in order to improve national preparedness. Moreover, the report states that the objective was not to predict the future, but instead to chart the potential development paths for the future. As the report states:

> The scenario descriptions are not meant to serve as definite predictions of the future. Instead, they are meant to steer versatile and consistent thought on potential development paths for the future. This is to improve the preparedness for interpreting current phenomena and bolster the planning of operations and ability to react. The idea is not to choose one scenario over the others on the basis of probability, desirability or importance, but to examine the whole formed by the scenarios. The actual future is often an amalgamation of different scenarios. [13]

It is thus a good example of a "fit for purpose" scenario process, where a number of diverse participants collaborate towards divergent potential future paths in order to assist in dealing with increasing uncertainty in long-term strategic planning. As the report also states, it is likely that none of the scenarios materializes as such, but instead the actual future may turn out to be an amalgamation of the scenarios. NESA is also currently working on a scenario planning projects focusing on the "new normal" of the post-COVID-19 world in order to determine how the pandemic may impact the objectives of its mission in the long-term.

Whilst best practices for applying scenario planning to security are scarce (this should be considered a call for action), the Centre for Risk Studies at the University of Cambridge has produced best practices for applying scenario planning for disaster risk reduction. This may well serve as a good starting point for security also. Another best practices guide, prepared for security and law enforcement agencies, by Centre for Research and Evidence on Security Threats (CREST) may serve as another starting point (CREST website).

5 Sensemaking: Answering "The So What?" and "What Next?"

In order to enable evidence-based decision-making in an increasingly complex, dynamic and uncertain world requires enhanced capabilities for making sense of the strategic environment in flux. One available measure is to improve the utility of *strategic intelligence* methods towards enabling such sense-making. *Strategic Intelligence* has no universal definition, but the term is generally made in reference to intelligence analysis that addresses the analysis of significant changes in the strategic environment that could have impact on the strategic objectives of the organization. In other words, its primary function is to enable strategic decision-making by addressing the uncertainties related to the organisation's long-term objectives. Unlike more operationally focused intelligence analysis, strategic intelligence analysis is "conceptual" [14, p. 5], rather than being focused towards any specific and individual threat, event or problem and responses to it. Thus, its objective is to create awareness about strategic change, its drivers and implications in relation to the organisation's strategy.

Whilst the history of strategic intelligence in the context of national security and defence is not entirely spotless and one could quote an impressive list of "intelligence failures" from the history—such as Pearl Harbour, the collapse of the Soviet Union (and the reasons behind it) or the 9/11, the focus here is not "intelligence failure" or "intelligence error" in the context of warning intelligence. Moreover, albeit such "intelligence failures" can be attributed to a number of potential root causes—for instance factual inaccuracies in analysis, insufficient data, failures of imagination and the failure to differentiate (weak) signals from the noise or a failure to "connect the dots"—it is equally possible that the intelligence is correct, but policy can still be a failure, either due to a failure to act on the intelligence provided, choosing the wrong measures despite what the intelligence would suggest or due to reacting to intelligence versus acting on it (failure to see the long-term implications) appropriately [15, pp. 12, 13].

Rather, the focus here is how intelligence analysis can enable decision-makers to cope with the vagueness and uncertainties associated with strategic uncertainty in the extremely challenging conditions dictated by the VUCA and TUNA dynamics of the external strategic environment. Consequently, decision- and policy-makers require the ability to understand the significance of the intelligence in the larger context of potential strategic-level change, either ongoing or developing, which may significantly impact making long-term decisions. This is of course primarily the policy-makers' responsibility, but it also means that traditional intelligence analysis, which is not necessarily specifically concerned with the broader context within which the problem exists (at least unless there is a detectable or clear causal relationship), can serve the needs of the VUCA/TUNA world. In this context, however, it is important to note that there are different types of problems intelligence may be asked to address, with great variance depending on the context in which intelligence work takes place (national/organisational, tactical/operational/strategic). These can be categorised according to the nature of the problem, as well as tasking, to "tame" or "wicked" problems. Of these two categories, "tame" problems are such in which there is a general agreement on the parameters of the problem and the intelligence targets/actors are clearly defined (or definable). Such problems are labelled as "tame", not because they would be by definition easy intelligence tasks (they can represent "puzzles" that are hard to solve nonetheless), but principally because the selection of solutions or policy-choices to address or mitigate the negative impacts of the problem are also relatively obvious [15, p. 17]. Consequently, in order to match the problem and the solutions, the intelligence analyst can build hypotheses that can be tested and verified. As such, the "pieces on the table" can be coherently solved into a puzzle and potential solutions and their impacts tested [15, p. 17]. The so-called "wicked problems",[4] however, are an entirely different game altogether as they specifically directly challenge such assumptions and approaches. First of all, by definition they are problems that have no clear parameter and, as such are not easily definable. This means that there is no clear endpoint to the problem that can be the focal point of the analysis. Furthermore, because of the lack of parameters of such problems the causal chains are also extremely hard to establish. Hence, the setting of a focus to the analysis and to formulate hypotheses and assumptions are also hard. Consequently, formulating an intelligence tasking around a "wicked problem" is challenging at the least. Perhaps even more challenging characteristics of the "wicked problems" is that, due to their undefinable nature, attempts to solve the problem may merely change the problem instead of solving it. Hence, it is also hard to justify an intelligence tasking that would expect the analysis to look for definitive solutions to a problem that is un-definitive by nature. There are no yes/no box-ticking options in response to a wicked problem, nor are there right or wrong answers. Instead, it is quite possible that there are many truths about the problem, its various causes and possible implications coexist and take precedence in the course of time. If the attempts to form workable hypotheses to understand the problems and potential solutions is nearly impossible, are "wicked problems" unsolvable? Not necessarily. Whilst they may challenge the traditional approach to increase understanding by increasing data and applying probabilistic statistical analysis, the nature of "wicked problems" demands a different mindset. This may include changing the course of action from attempting to control or manage the problem to adapting to it. Even before COVID-19 pandemics have been utilised as an example of "wicked problems" [15, p. 29]. However, COVID-19 certainly seems to have the potential to enforce the "wickedness" of pandemics to a new and unexplored extent. Unlike the previous post-WWII pandemics, such as the "bird flu" and "swine flu", COVID-19 appears to be the first pandemic after the "Spanish Flu" in 1918–1920 that defies the attempts to contain it due to its adaptive and recurring nature. It is this adaptive and recurring characteristic of COVID-19 that makes it possible that we will suffer a second wave, which may be followed by a third, fourth or even a fifth wave. Moreover, whilst it will at some point come to its end, there is significant uncertainty in regard to whether vaccines, anti-viral medicines and other mitigation measures—such as quarantines, border closures, testing or social distancing—will eventually be sufficient to exhaust the expansive potential of the pathogen. In addition to the uncertainty relating to the effectiveness of mitigation measures, we have still significant uncertainty about the direct and indirect consequences of the mitigation measures. Some of these impacts have of course started to emerge already, e.g. the economic impact from quarantines and border closures is expected to become enormous. At the time of the writing this effect is already mounting up. The International Monetary Fund (IMF) has estimated the global economy will shrink by 3% in 2020, making it the worst decline of global

---

[4]The concept of "wicked problems" was first introduced by Horst Rittel and Melvin Webber [16] in their 1973 article "Dilemmas in general theory of planning" in *Policy Science* 4(2).

economic activity since the Great Depression in the 1930s (BBC News). The World Bank, in its *Global Economic Prospects*—June 2020, on the other hand, forecasted a 5.2% contraction in global GDP in 2020, with most of the countries risking recession (World Bank). Moreover, it points out to significant uncertainty and downside risks in the global economic outlook with prospects for much worse if the assumption of the pandemic receding mid-year would not hold and restrictions are re-introduced around the world. In this case the downside scenario could signify the global economy shrinking by 8% in 2020. Furthermore, there are systemic level risks looming that may become the cause of a major long-term economic downturn—ranging from crash in demand of oil and prices, decreasing levels of investment and industrial production, erosion of human capital, rising and systemic unemployment and frag-mentation of global trade and supply chains (World Bank, UNIDO). Whether, we will experience something still much worse largely depends on the future path of the pandemic and the measures required for mitigating the human security impacts. This interlinked dynamic is emerging in data, where the economic impacts correlate with health impacts (UNIDO). Hence, the "wickedness"—is the difficult trade-off between health security and economic interests. The more severe the health control measures are the more severe is the impact on the economy. Moreover, the longer this dynamic continues the less sustainable it is. At the same time there is no evidence, as of yet, that there is a correlation between the number of COVID-19 related deaths and industrial output (UNIDO), further complicating the long-term prognosis and strategic options to the crisis.

Despite all this, the long-term economic impact is nonetheless still somewhat unpredictable, as in addition to the impacts materialising with delay, they are not only on the downside but possibly also offer some upside opportunities. Whilst it is rather clear that many business have suffered extensively from quarantines and border closures, and that some of them will go out of business and fail altogether, whilst others have to take measures to reduce cost and retrench people, the result is an immense economic impact that will place further strain on national economies. However, good things can happen as well. First of all, a number of countries and industries have experienced a digital leap as a positive reaction to restrictions on human contacts. This has not only resulted in the exponential growth of remote working, but also increasingly the conversion of services that have traditionally been delivered through some form of physical contact into digitally delivered. The result has been an efficiency boost and an abundance of new innovations, leading to a disruptive renewal from the crisis that will likely create also new business.

Hence, we may have to find a way to adapt to COVID-19, i.e. learn to live with it, instead of trying to contain, exhaust or mitigate it. This would also entail the ability to look past the current crisis and prepare for the post-Covid "new normal", whatever that may be. The point being that whilst we have the ability to address individual problems in connection with the pandemic, such as innovating new methods for testing and vaccines that help curtailing the spread, we do not have the ability to determine what the end game will look like. Hence, we are unable to draw the parameters around the problem in order to solve. Moreover, the problem is larger than the pandemic and we still do not know what the indirect and compound impacts will be. In sum, COVID-19 may prove to be unsolvable, but not un-survivable, i.e. a true "wicked problem".

What the COVID-19 pandemic does tell us, is that we are not prepared for wicked problems with the magnitude of a pandemic. How will we be able to spot "black swans" (albeit COVID-19 was not one) that may become "wicked problems" early enough and enable adequate responses in time? How will we ensure we will not be surprised again? This is where sense-making steps into the picture.

5.1 Strategic Intelligence and Sensemaking

Since "strategic intelligence" by definition focuses on supporting decision-makers in dealing with strategic level ambiguity, it focuses primarily on long-term trends and other macro-level changes. There is, however, increasing levels of uncertainty associated with increasing scales of complexity and ambiguity that comes with such change. Hence, there are no right answers or responses that have a high-level of certainty associated with them. When there is a lack of immediate responses available and the causal chains between various change drivers are hard or impossible to establish, the challenge probably qualifies as a "wicked problem". Hence, strategic intelligence should enable decision-makers to deal with such problems by offering means for increasing contextual understanding in order to seed the meaningful from the noise. This essentially boils down to the ability to make sense of a situation that defines all "quick fixes" and requires a more sustained strategic response. This capability has been titled as "sensemaking". Sensemaking, a broad and largely undefined concept first introduced by Weick [17], has been suggested to mean a process through which one becomes aware of "something, in an ongoing flow of events, something in the form of a surprise, a discrepant set of cues, [or] something that does not fit" [17, p. 2]. In the simplest of terms, Weick describes the process as "the making of sense" [17, p. 4]. By "sensemaking" as an activity he explains that sensemaking enables leader to "explore the wider system, create a map of that system, and act in the system. This in turn enables leaders and organisations to "comprehend, understand, explain, attribute, extrapolate, and predict" the unknown. In other words, it is an activity and ability that helps us to structure the unknown in a manner that enables us to act to it [18, p. 3]. Significantly for the theme of this chapter, sensemaking can be useful for establishing how much (or little) we know about the unknown [18, p. 4]. Thus, it is a viable option for addressing the "unknown-unknowns" problem coined by Donald Rumsfeld, where understanding cannot be increased by more data, especially since there might not be any specific data available. Moreover, sensemaking can equip the individual or organisation with the ability to anticipate uncertainty, instead of merely reacting to it. In order to work effectively, the process of sense-making should be detached from the "heat of the moment", as well as the status quo and conventional thinking normally associated with policy-making [15, p. 9]. If applied properly, sensemaking can create awareness and sensitivity to rapid and unexpected change in a more dynamic, complex and uncertain world. It is particu-larly useful in dealing with issues with high levels of ambiguity and novelty and as such can be useful for enabling preparedness against still unknown threats. As such, it should be at the core of forward-looking strategic intelligence.

6 Closing the Loop: From Detection to Action

Whilst *extreme (deep) uncertainty* is primarily, in the context of the future is an exogenous condition beyond our direct control and thus, something that cannot be eliminated entirely, it can be reduced, principally by establishing and developing the organisation's situational awareness. This essentially entails applying methods that increase the organisation's capabilities to identify, understand and assess both sudden and gradual changes in its strategic operating environment. In other words, the organisation should aim at establishing processes for successful anticipation of significant trends early enough to build resilience against and adapt to negative change strategically, as well as more constant and continuous measures to establish awareness that enables taking rapid action (operational measures) in face of sudden shocks in order to ensure the survival of the organisation. The first line of defence thus, is reducing the probability of being surprised of shocks or disruptive changes that result from long-term trends can be reduced by investing in "strategic foresight", which entails a scenario building approach towards identifying and understanding possible futures in order to build strategies both for exploiting the opportunities and insulating the organisation against the threats emanating from them. The second line of defence focuses on monitoring of trends and active horizon scanning to spot weak signals that may indicate more sudden significant changes in the operating environment prior to them having an impact on the organisation. This enables the organisation to take measures for adaption and defences before *creeping crises* erupt.

In order to reduce the number of occasions and the impact of getting taken by surprise, the organisation need to have the capability to continuously observe its operational environment for signs of potentially significant changes, both long-term and sudden. This entails two critical capabilities; sensing the relevant changes and making sense of them.

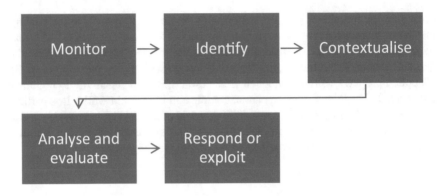

The flow from detection to action. *Source* Author

Finally, it should entail an ongoing function of situational awareness, which is also connected with incident reporting across the organisation for early detection of events that may be already impacting the organisation. This enables the organisation to take rapid corrective action in face of an unexpected and rapidly escalating situation, by depending on the scale of the situation activating its incident-, major event or crisis management procedures. In order to establish such situational awareness an organisation should have the capability to **monitor** its operating environment in order to **identify** changes that may be relevant to its operations. The organisation also needs the capability to **contextualise** such changes in order to understand and make sense of their meaning to its particular circumstances and to subsequently **analyse and evaluate** the likely impacts on its operations. This provides the organisation the decision-making basis for deciding whether to **respond** to the event in order to limit its negative impact, or to **exploit** the circumstance created by the event in order to leverage on its relevant competitive strengths. This is the path towards being caught by surprise less often and less prepared.

7　Conclusion: The Path to Being Surprised Less Frequently

The contemporary strategic environment is more complex, dynamic and uncertain than perhaps at any point in history. Consequently, the threats of both sudden crises and creeping crises test our ability to take action before the negative impacts materialise. In order to be prepared against the perils of deep uncertainty we need to collectively improve our capabilities to monitor the strategic environment, identify relevant changes, make sense of them, monitor and prepare for them proactively and sufficiently. This chapter has explored some of the available methods and approaches for achieving this, but there is still much to explore. We simply need to start taking the deep uncertainty more seriously and seek ways to improve our preparedness against it. COVID-19 pandemic is a testimony of the fact that our current capabilities are not adequate, nor is it acceptable to be satisfied with the status quo and the limited utility of probabilistic risk management methods and traditional intelligence in dealing with the unknown. We need a revolution of capabilities, not a slow evolution.

References

1. Global Preparedness Monitoring Board (2019) A world at risk: annual report on global preparedness for health emergencies. The World Health Organization, Geneva. https://apps. who.int/gpmb/assets/annual_report/GPMB_Annual_Report_English.pdf
2. Taleb NN (2007) The black swan: the impact of the highly improbable. Random House, New York
3. Phillips F (2020) What about the future? New perspectives on planning, forecasting and complexity. Springer
4. Goldin I, Mariathasan M (2014) The butterfly defect: how globalization creates systemic risks, and what to do about it. Princeton University Press, Princeton
5. Aaltola M, Käpylä J, Mikkola H, Behr T (2014) Towards the geopolitics of flows: implications for Finland. FIIA report 40. The Finnish Institute of International Affairs (FIIA), Helsinki
6. Fjäder C (2016) National security in a hyper-connected world. In: Masys A (ed) Exploring the security landscape: non-traditional security challenges. Springer
7. Fjäder C (2018) Interdependency as dependency: economic security in the age of global inter-connectedness. In: Wigell M, Scholvin S, Aaltola M (eds) Geo-economics and power politics in the 21st century: the revival of economic statecraft. Routledge, London
8. International Standards Organisation (2018) ISO 3001: risk management. Geneva. Available from www.iso.org
9. Knight F (1921) Risk, uncertainty and profit. Houghton Mifflin Company, Boston
10. Jarvis D (2011) Theorising risk and uncertainty in international relations: the contributions of Frank Knight. Int Relat 25:296–312
11. Ramirez R, Wilkinson A (2018) Strategic reframing: the Oxford scenario planning approach. Oxford University Press, Oxford
12. PreventionWeb (2020) Interview: scenarios, an effective tool to understand and mitigate risk, 26 Mar. https://www.preventionweb.net/experts/oped/view/71050
13. National Emergency Supply Agency (2017) Security of supply: scenarios 2030, Helsinki. https://cdn.huoltovarmuuskeskus.fi/app/uploads/2018/09/06091431/Eng-Scenarios-2030.pdf
14. McDowell D (2009) Strategic intelligence: a handbook for practioners, managers and users. The Scarecrow Press Inc, Lanham
15. Moore DT (2011) Sensemaking: a structure for an intelligence revolution. In: National Defense Intelligence College (NDIC), Clift series on the intelligence profession. NDIC Press, Washington. https://ni-u.edu/ni_press/pdf/Sensemaking.pdf
16. Rittel HWJ, Webber MM (1973) Dilemmas in a general theory of planning. Policy Sci 4(2):155–169
17. Weick KE (1995) Sensemaking in organizations. Sag, Thousand Oaks
18. Ancona DL (2011) Handbook of teaching leadership. Sage, Thousand Oaks, pp 3–20

Global Health and Pandemics—Beyond Direct Effects of COVID 19 Outbreak

Sara Spowart and Anthony J. Masys

Abstract The COVID 19 pandemic has resulted in significant national and global public health impact. The mortality and morbidity statistics associated with COVID-19 has become a key impact metric. As of 26 June 2020, globally upwards of 10 million cases of COVID 19 have been reported and 500,000 deaths. In the US alone there have been upwards of 2.5 million cases and 125,000 deaths. The EU/EEA and UK have reported 1.5 million cases and 176,000 deaths. The viral effect associated with COVID 19 is well tracked stemming from the direct effects of the virus. However, there are secondary and tertiary public health effects stemming from the COVID 19 pandemic that requires attention and management. These include: psychological distress, loneliness, mental health issues, and domestic violence. Developing public health strategies associated with the COVID-19 pandemic requires sensitivity to unexpected outcomes, to sensemaking that explores beyond the direct effects but considers the second and third order effects. This chapter will examine these secondary and tertiary public health effects and will highlight the requirement for improved public health surveillance and awareness regarding these effects.

Keywords COVID 19 · Pandemic · Domestic violence · Mental health · Vulnerable populations

1 Introduction

The COVID 19 pandemic has resulted in significant national and global public health impact. The mortality and morbidity statistics associated with COVID-19 has become a key impact metric. As of 26 June 2020, globally upwards of 10 million cases of COVID 19 have been reported and 500,000 deaths. In the US alone there have been upwards of 2.5 million cases and 125,000 deaths. The EU/EEA and UK have reported

S. Spowart · A. J. Masys (✉)
College of Public Health, University of South Florida, Tampa, FL, USA
e-mail: Anthony.masys@gmail.com

A. J. Masys
International Centre for Policing and Security, University of South Wales, Pontypridd, UK

1.5 million cases and 176,000 deaths. The viral effect associated with COVID 19 is well tracked stemming from the direct effects of the virus. However, there are secondary and tertiary public health effects stemming from the COVID 19 pandemic that requires attention and management. These include: psychological distress, loneliness, mental health issues, and domestic violence. Developing public health strategies associated with the COVID-19 pandemic requires sensemaking that explores beyond the direct effects but considers the second and third order effects.

For example, the opioid crisis around the world has been exacerbated by the COVID-19 pandemic. As described in ([1]: 5):

COVID-19 has caused the worst economic devastation in decades, potentially increasing susceptibility to drug use and OUD and restricting the availability of treatment. COVID-19 further exacerbates the physical, emotional, social, and economic challenges for OUD sufferers and their families and communities. A source of massive economic hardship and physical and emotional pain in of itself, COVID-19 and the economic dislocation associated with necessary lockdowns may increase mental illness—augmenting OUD susceptibility and associated overdose risks. According to a recent study, COVID-19 could produce 75,000 deaths of despair, including from suicide and drug overdose, in the United States alone.

The Covid-19 outbreak has resulted in considerable impact on global health security, human security and has also impacted the global economy. As described in ([2]: 2) 'the pandemic has reminded us bluntly of the fragility of some of our most basic human-made systems. Shortages of masks, tests, ventilators and other essential items have left frontline workers and the general population dangerously exposed to the disease itself. At a wider level, we have witnessed the cascading collapse of entire production, financial, and transportation systems, due to a vicious combination of supply and demand shocks'.

This certainly connects with Weick and Sutcliffe [3] reflection: '...unexpected events often audit our resilience. Everything that was left unprepared becomes a complex problem'. Taleb [4] argues that 'not seeing a tsunami or an economic event coming is excusable, building something fragile to them is not'. This points to the importance of sensemaking across the COVID-19 pandemic disaster/crisis management cycle (Mitigation, Preparedness, Response, Recovery). The COVID 19 outbreak along with other public health crisis over the last few years has characterized the "new normal" unveiling major deficiencies in preparedness, response and recovery initiatives locally, regionally and globally.

A systems thinking perspective on COVID-19 brings to the forefront the inherent interdependencies and interconnectivity that exists across national and global societal health infrastructure revealing vulnerabilities for cascading second and third order effects. This systems lens opens opportunities for shaping risk perceptions beyond direct effects to enable better crisis/disaster management and community resilience.

1.1 Systems Thinking

We are dealing with a world characterized by nonlinearities, tipping points, and complex relations where the 'butterfly effect' is a constant reminder of our need for sensemaking at scale. In a systems approach, global issues need global solutions. As with the COVID-19 crisis, a disease anywhere is a disease everywhere. Disease outbreaks do not respect national borders.

Massaro et al. [5] argues that 'The evaluation of vulnerabilities and consequences of epidemics is a highly dimensional complex problem that should consider societal issues such as infrastructures and services disruption, forgone output, inflated prices, crisis-induced fiscal deficits and poverty. Therefore, it is important to broaden the model-based approach to epidemic analysis, expanding the purview by including measures able to assess the system resilience, i.e. response of the entire system to disturbances, their aftermath, the outcome of mitigation as well as the system's recovery and retention of functionality'. Systems thinking figures prominently in the discourse pertaining to disaster aetiology and disaster forensics [6]. Jackson ([7]: 65) defines systems thinking paradigm as'…a discipline for seeing the 'structures' that underlie complex situations, and for discerning high from low leverage change…Ultimately, it simplifies life by helping us to see the deeper patterns lying beneath the events and the details'. Systems thinking also provides an approach to challenge our inherent beliefs and assumptions regarding causality and the impact of intervention strategies.

As described in Ramos et al. ([8]: 3), 'Systems thinking not only improves multi-disciplinary, cross-sectoral collaboration, it can also provide insights into systems behaviour and management by rigorous analysis of such aspects as system dynamics, feedback, sensitivity, and non-linear responses,the emergence of systems behaviour and properties; the optimisation of system performance over different time horizons or for different groups; the anticipation and assessment of systemic risks; and the strengthening of resilience to external change and shocks'. Given the complex aetiology associated with the emergence and impact of COVID-19 on global health security, systems thinking emerges as a key sensemaking approach.

2 Sensemaking

COVID-19, aside from the viral implications, triggered an economic crisis with significant global societal consequences. As noted in WEF ([9]: 8),

> The crisis has exposed fundamental shortcomings in pandemic preparedness, socio-economic safety nets and global cooperation. Governments and businesses have struggled to address compounding repercussions in the form of workforce challenges, disruptions in essential supplies and social instability. They have had to balance health security imperatives against the economic fallout and rising societal anxieties, while relying on digital infrastructure in unprecedented ways. As countries seek to recover, some of the more lasting economic, environmental, societal and technological challenges and opportunities are only beginning

to become visible. While societies, governments and businesses collectively grapple with these possibilities, it is vital to anticipate the emerging risks generated by the repercussions from the pandemic.

Given the societal effects of COVID-19, sensemaking and risk perception emerge as essential elements in managing the pandemic. Weick [10] refers to sensemaking in terms of '…to how we structure the unknown so as to be able to act in it. Sensemaking involves coming up with a plausible understanding—a map—of a shifting world, testing this map with others through data collection, action, and conversation; and then refining, or abandoning, the map depending on how credible it is' [11]. Ancona [11] describes sensemaking as an '… activity that enables us to turn the ongoing complexity of the world into a "situation that is comprehended explicitly in words and that serves as a springboard into action". Thus sensemaking involves— and indeed requires—an articulation of the unknown'. Making sense of a 'complex' world and all that characterizes 'complexity' is a key theme in the sensemaking and crisis leadership literature.

3   Discussion

> It was the worst experience of my life, I've never been so sick. It took me more than 3 months to recover and I'm still not completely better. It feels like your lungs are shutting down and are burning for weeks and weeks. I hope no one goes through what I went through, it feels like you are dying and are afraid you'll never recover. It affects your brain, how food tastes… it feels like it's the end of your life.
>
> —Anonymous statement from a COVID-19 survivor recovering from a severe case of COVID-19

3.1   Overview of COVID-19

COVID has become the pandemic of our time and worldwide it has spread to become the fastest rising health issue in the world. It is currently in route to become one of the top 10 causes of death worldwide. COVID is a highly infectious disease that has not only severe ramifications and health concerns for those that become infected but additional ramifications that cause secondary and tertiary effects for businesses and the economy, as well as other psychological issues and concerns such as depression, suicidal ideation, anxiety, loneliness, domestic violence, unemployment, and other health concerns in relation to COVID such as not being able to obtain necessary medical treatment for other health concerns that are deemed less urgent [12].

For example, heart surgeries that have been viewed as elective and other medical procedures have been postponed or canceled because of reductions in staff and COVID precautions are not only impacting the treatment of COVID but also individuals with other health conditions. There is a rise of physical and mental health conditions, particularly depression and feelings of hopelessness in many parts of the world. There is also something increasingly occurring termed 'quarantine fatigue' and refers to issues that accompany COVID. This can be feelings of exhaustion at being home all the time to shelter in place, also the stress and continual uncertainty of what is happening in the world and in the future. The feeling of uncertainty and chronic stress of not knowing what to expect is potentially the most challenging aspect of COVID. There is no certainty of when the pandemic will end or if it will end. The entire world is having to adapt to a new kind of normal, a normal that is filled with fear of disease and poverty and accompanying issues that coincide with this [12].

The impacts of COVID-19 infection are not only the immediate symptoms of the disease which include such issues as fatigue, cough, respiratory issues, pneumonia, shortness of breath, muscle pain, confusion and more. One of the primary impacts of COVID-19 is a high level of inflammation in the body. This is due to an overreaction of the immune response. This inflammatory response can lead to inflammation in the heart and muscle lining, inflammation of the lungs and even stroke. Recent medical interventions that have proven lower cost and effective for severe cases of COVID-19 include the use of the steroid Dexamethasone to reduce the body's inflammatory response and its harmful consequences [13].

3.2   Examination of Secondary and Tertiary Effects

As a result of the coronavirus, many elective surgeries have had to be adjusted and postponed. On March 11, 2020 the World Health Organization declared SARS-CoV-2 a global pandemic. It was initially reported in the Hubei region of China. Some of the first areas to be affected internationally were the Veneto region of north-eastern Italy. The first cases to be reported in Veneto, Italy were reported on February 21, 2020. On April 27, 2020 the Veneto region had 17,579 confirmed SARS-CoV-2 cases with 1099 hospitalized patients and 1344 deaths [14].

Aside from COVID reducing the number of elective surgeries that are permissible due to COVID capacity, it has also reduced the number of hospital beds and available health care workers for other health concerns. Other health care issues that are of critical importance but could be pushed to elective medical concerns include elective heart surgeries, diabetes, dentistry concerns, gynecological issues, and any issues that are considered elective or can be pushed to a later time. The reason for this is to allow enough space in ICU beds and hospitals so that severely sick COVID patients are able to access desperately needed resources [14].

However, as a result of this, individuals with chronic health conditions or new and acute health conditions that need assistance may not be able to access the healthcare they need. But at the same time the current situation with COVID is also allowing for new and innovative ways to address healthcare concerns. For example, telehealth is allowing for changes in the ways that individuals can access healthcare providers so that it is no longer necessary to leave one's home when they need to access help and necessary prescriptions can be called in by doctors through telehealth services. Therefore, for those with access to internet and video platforms, there may be better healthcare support and fewer barriers such as reduced transportation concerns and time constraints. Also, it provides reduced chances to be exposed to COVID by remaining at home and also not exposing your immune system to additional concerns when it is already compromised [14].

Additional effects include the impact on the economy. Specifically, COVID has led to a reduction in production of products and supplies. Factories and suppliers have had to close or severely limit their operations to reduce the number of COVID infections. Certain countries have been hit harder than others in terms of production. For example, Chinese production has been severely impacted by the closures that were necessary in some regions of the country such as the Hubei province and other nearby areas. These shut downs also impact the exports coming into China for production. The biggest exporters to China are Korea, Japan and other Asian countries. The slowdowns have created disruption in supply chains and in marketplace functioning. These disruptions are impacting small and medium size firms the most and they are least likely to survive the economic downturns [14, 15].

COVID-19 has led to numerous secondary and tertiary effects. A primary effect is caused directly by the disaster or crisis. In the case of COVID-19 this refers to the actual infection from the virus. Secondary effects are what occurs as a result of the primary effect. The secondary effects of being infected with COVID-19 include fatigue, respiratory issues, stroke, fever, cough, headache, muscle weakness and pain, heart disease, inflammation, concentration issues and difficulty focusing, hallucinations, loss of taste and smell, and a seemingly growing list of symptoms that are surprising healthcare workers. Some of the newer symptoms on the list include anxiety, depression, and even Post Traumatic Stress Syndrome and symptoms associated with this condition. There is increasing evidence of not only the many physical symptoms from the virus but the accompanying mental health conditions due to the virus [14, 15].

COVID-19 is a disease that takes many shapes and forms ranging from asymptomatic, mild, and severe physical and mental health conditions. Tertiary effects are the long-term effects and impact that is created as a result of the primary event. In the case of COVID-19, not only are we globally experiencing significant impacts from primary and secondary effects, but the tertiary effects appear to be mounting in depth and extent and there doesn't appear to be an end in sight to the interconnection and domino-like effects occurring as a result of COVID-19. Some of these include the massive global economic losses and marketplace disruption worldwide. These also include the untreated mental and physical health conditions that are not able to be treated due to the burden to many countries healthcare systems [14, 15].

Additional tertiary effects include the massive job loss and unemployment that is impacting not only the United States but countries worldwide. Due to the interconnected nature of the current world economy, now more than ever in the world's history what occurs economically to one country or region impacts other countries and parts of the world. Other tertiary effects are the isolation, loneliness and accompanying depression many are experiencing as a result of social distancing, quarantine and lock-down, working remotely and from home, inability to travel and other forms of isolation. In addition to this, worldwide there has been a significant increase in domestic violence rates [16].

3.2.1 Domestic Violence

We had a COVID-19 outbreak at the shelter. It was very difficult. We had to have the women and children stay in a hotel while we had to sanitize everything. But now we have to reduce our capacity at the shelter. We won't be able to take in as many women and children to get away from their abusers now because of COVID.

Because of the COVID outbreak, staff and clients are more scared for their health. But there are also a lot of people we can't take to help because of COVID. It's really sad what is happening and what we have to do to keep everyone safe at the shelter.

—Anonymous statement from a domestic violence shelter manager on the state of the shelter and its residents

Domestic violence is actually deadlier and a greater health risk for women in many countries than COVID due to the lockdowns. While the lockdowns have potentially saved the lives of many, it has put countless individuals at greater risk of domestic violence and intimate partner violence worldwide. To be forced to be with an abusive individual, the words "lockdown", "self-isolate," "stay at home" "recession" and "practice social distancing" are a nightmare and potential death threat for individuals living with intimate partner violence. In Mexico alone, more individuals have died from intimate partner violence than from COVID. Before the COVID pandemic, it was estimated that 1 in 3 women worldwide experienced intimate partner violence and 1 in 4 experienced severe intimate partner violence. Other data show that half of all female homicides are from a current or past intimate partner. However, despite these numbers being very high, this data reflects general rates that occur. Rates of intimate partner violence, severe intimate partner violence and homicide related to domestic violence increase when families spend more time together such as holidays, when there are pandemics and economic crisis [17].

Due to COVID there have been significant increases in hotline calls requesting assistance for intimate partner violence, with some regions experiencing double or triple the rates of hotline calls. Conversely, some areas have experienced an eerie silence with hotline calls and a significant and alarming decrease in response. Evidence suggests that areas with decreases in response are due to intimate partner violence victims not feeling safe enough or being unable to make calls for help. The rates of femicide being witnessed worldwide in relation to lockdowns and COVID-related economic crisis reflect this. For example, in the United Kingdom, femicide rates were higher than they have been in the last 11 years and twice as high as usual over a 21-day period. In Mexico, more than 1000 women were murdered in only the first three months of 2020, this is an 8% increase in femicide rates. The escalation in intimate partner violence also points to more severe forms of violence that are occurring related to economic, quarantine and COVID stressors. The increase in severe intimate partner violence is not only creating an increase in femicide rates.

It is also resulting in higher rates of traumatic brain injuries (TBI's) and a greater number of repetitive traumatic brain injuries. Traumatic brain injuries lead to cognitive, psychological and neurological difficulties that make it even harder to leave. Also, a woman's ability to leave is threatened because of quarantine and lockdown it can be nearly impossible to leave or seek help. Also, the closure of courts due to COVID has made it much more difficult or impossible for women to obtain order for protection or women that were planning to leave are no longer able to due to economic changes and COVID and fear of getting the virus or giving it to their children [17–19]

However, even despite COVID, by adopting a public health approach where everyone is responsible for looking out for each other and paying attention to someone who is very isolated or potentially experiencing abuse, it is possible to reduce the amount of risk and improve access to help and resources for victims. Specifically, if there is someone you know who is suffering from abuse or wanting to get out of relationship situation, it is helpful to let them know that you are available if they need help, to offer a place to stay if you are able, and check in often to see how they are doing so they know they have support and are not alone. Medical health professionals can assist by informing patients that intimate partner violence and domestic violence rates have increased since the outbreak of COVID and to check if they feel safe at home and provide resources such as hotline information to provide help and discuss a safety plan [20].

3.2.2 Loneliness

Another component of COVID is the social isolation that has occurred as a result of the quarantine, lockdowns, social distancing, decreased employment, economic issues and relationship concerns. Social isolation is a significant component of morbidity and early mortality. Social isolation and loneliness were prevalent worldwide before COVID, however research points to the increase in these rates as a matter of increased public health concern since COVID. Social connection reduces the risk of premature mortality whereas social isolation increases the risk of early mortality. Social isolation is associated with increased morbidity of significant health concerns such as inflammation [21–23].

The accompanying social isolation epidemic in connection with COVID is creating a need for a greater number of evidence-based interventions and research to reduce social isolation. Social isolation refers to an objective state with few or scarce social contacts. Loneliness refers to the subjective state and emotional pain caused by social isolation. Loneliness is the gap between desired and actual level of social interaction. Strong social connection is a protective factor to health. Loneliness is on the rise nationwide and by some estimates, even before COVID was considered a public health epidemic. In 2020, Cigna conducted a study and reported that loneliness was up from 47% of the population in 2018 to 61% in 2019. Mental health and physical health problems, living far from family or living alone can increase the potential of loneliness and social isolation. Loneliness is found in all age groups and backgrounds. Poorer social connection is associated with decreased social wellbeing and lower immunity. It is even associated with an increased chance of developing the common cold. Currently, there are national health guidelines on lifestyle aspects that contribute to health. Some of these include nutrition, exercise and sleep for example. However, due to increasing evidence that social connection and isolation negatively impact health outcomes, there is an increasing amount of data to show that these should also be incorporated into national guidelines. Like guidelines on nutrition, sleep and exercise, these recommendations should provide information on frequency of social connection, such as daily, or twice a week and type of social connection, such as phone, in person, video, etc. [21–23].

With COVID-19 there has been a significant increase in the use of phone and video technology for social connection and interaction. There are reports of weddings and important events all being conducted by Zoom, conferences and social networking being conducted by platforms such as Microsoft Teams and webinars. Therefore, although social connection and interaction may not be in-person as frequently, it can be substituted and potentially increased with the use of greater access to technology. Unfortunately, although many both in the United States and worldwide have much greater access to technology than ever before in history, it can still be limiting, intimidating or potentially isolating [21–23].

This is especially true for certain individuals that prefer in-person interaction. There are often stereotypes that certain groups or "types" of people are better adjusted to the use of technology to decrease the experience of social disconnection and isolation than others. For example, the idea or belief that children of a certain age group or demographic are much better adapted to technology than individuals that are 55+ years. However, it often depends on the individual and their personal perspective on technology as well as their motivation and needs. No two individuals are necessarily the same in their view on the use of technology and it can be discriminatory to make assumptions and generalizations. In addition to this, it can be presumptuous to think that just because an individual is elderly and living alone that they are lonely. Loneliness is a subjective experience and must be assessed using subjective measures for data collection. More research is needed to assess the dynamics and on-going adaptation and experience of loneliness and perception of social disconnection during this time with COVID. In order to provide the best possible health guidelines and recommendations to manage the impact of loneliness and social disconnection on physical and mental health outcomes are required [21–23].

3.3 *Psychological Distress*

Mental health status of individuals in every socio-economic background, gender, ethnicity and culture worldwide are impacted negatively by traumatic events and the negative consequences of social and economic changes. Large scale traumatic events have historically demonstrated significant negative impacts particularly as seen in historical disasters, epidemics and civil unrest. Furthermore, mental health impacts are not experienced evenly across a population. The negative impacts can generally be experienced across an entire population; however, lower incomes groups tend to be associated with greater burdens of mental illness in times of trauma and situations such as epidemics, pandemics or civil unrest. Currently, there is not only the impact of COVID-19 on mental health, but there are also the accompanying impacts of civil unrest connected to the Black Lives Matter movement as well as associated low employment, decreased income and decreased sense of security in the present situation and future. For example, the 2019 Hong Kong Civil protests demonstrated higher levels of depression, anxiety and psychological distress in association with the civil unrest and violence. Civil unrest, protests and violence alone create higher levels of psychological distress. This is worsened by large scale disasters such as pandemics [24].

Increases in psychological distress have also impacted healthcare workers. In addition to this, healthcare workers exposed to COVID-19 through patients demonstrate increased levels of depression and anxiety and decreased levels of sleep. In a study on health care workers, 50.4% of health care workers in China exposed to patients with COVID-19 reported new symptoms of depression. However, in general there remains a gap in the literature and understanding on the incidence and prevalence of mental health and mental health related concerns in the United States and worldwide during COVID-19. Yet, despite this gap in understanding, there is a projection that greater mental health support will be needed globally in the future post-COVID 19 [24].

A survey conducted by the John Hopkins Bloomberg School of Public Health during the COVID pandemic found that psychological distress has increased more than three-fold from 3.9% in 2018 to 13.6% in April 2020. The percentage of adults experiencing psychological distress in the United States ages 18–29 years increased from 3.7% in 2018 to 24% in 2020. It is believed that the social disruptions due to the Coronavirus outbreak, social distancing, economic uncertainty, high unemployment, and fear of contracting the disease are all negatively impacting mental health rates in the United States. Aside from age group, household income was also a significant determining factor in the increase of psychological distress rates from 2018 to 2020. This survey also found that adults with household incomes less than $35,000 a year experienced a significant increase in psychological distress from 7.9% in 2018 to 19.3% in 2020. In addition to this, psychological distress rates doubled in adults age 55+ from 2018 to 2020. Interestingly, according to this survey there was only a slight increase in rates of loneliness from 2018 to 2020, suggesting that loneliness is not a significant factor in the increased psychological distress [14, 25, 26].

By contrast, according to the Center for Disease Control in the United States during late June, 2020, 40% of U.S. adults reported struggling with mental health issues or substance use. Approximately 31% of adults reported struggling with anxiety and/or depression symptoms. In addition to this, approximately 26% of adults reported suffering from trauma and/or stress related disorder symptoms. Also, 13%

of adults reported that they started or increased substance use and 11% have seriously considered suicide [27].

3.4 Mental Health Issues

3.4.1 Suicidal Ideation

> Most teenagers aren't addicted to technology; if anything, they are addicted to each other.
> —Danah Boyd.

Suicidal ideation is a problem among teenagers in the United States today, and this is something that has only been worsened by COVID-19. There have been steady increases in suicidal ideation with a significant increase since 2018. However, suicide cannot be viewed as a something that largely or only affects certain socioeconomic backgrounds, ethnicities, age groups or genders. It is an issue that can impact every age group and individual. This is significant because generalizations or stereotypes that it mostly impacts the elderly or Caucasian individuals, for example, takes away from the inherent issue and concern at hand as well as better investigating and navigating triggers and contributing causes. Specifically, there are individual, family, community/provider, state/national influences on suicide outcomes. In the past, the most common approach has been to call the police and request emergency 911 services to dispatch to the crisis scene. The police often arrive on scene and may need to restrain or handcuff the individual and then take them to the hospital to be sedated in emergency services. This prior way Baker-Acted individuals with suicidal ideation or a plan to commit suicide were treated by being admitted to a hospital emergency room, and then held in a Baker-Act facility for up to 72 h or possible longer. If an individual with severe mental illness was not demonstrating sufficient progress during the time of being Baker-Acted in the hospital, they can go on to be committed to a longer stay even extending several months or longer if they are deemed a danger to themselves or others [27].

However, one of the issues with this approach is that although hospitalization for mental illness does keep an individual physically safe and others safe from them, it does not treat the core issues of the mental health condition. Also, it is assuming that suicidality is a result of mental illness when in fact it could be due to other factors such as work, financial, family, health, loneliness and other life factors. It may not be due to an inherent mental health condition and our current approach works to treat it as such. In addition to this, being hospitalized is most useful as a form of stabilization and to reduce the urgent nature of the suicidal ideation crisis. However, given this, if suicidal ideation is due to factors such as family environment and family conflict, financial and work concerns, addiction or drug and alcohol abuse, physical health issues, or other life factors, hospitalization only works to manage these factors in the short term so the individual is unable to act on them. But, it does not solve the core concerns [28].

For individuals ages 10–19 years of age, the five most significant issues that co-occurred with suicide death were school problems, arguments or conflicts, intimate partner issues, family relationship problems, and a crisis in the last two weeks or anticipated in upcoming two weeks. The experience of a crisis in the prior two weeks or an anticipated crisis in the upcoming two weeks was the most significant issue for suicidal ideation and suicide attempt. All of these concerns collectively point to the significance of connection, support and extreme stress or feeling of lack of safety. It points to the possibility that the lower the amount of perceived safety, the higher the amount of stress and decrease in feelings of control. Also, other significant aspects are that approximately half of suicides for individuals ages 10–19 years identify as gay or homosexual, which indicates a considerable relationship for public health concerns and a point of intervention [28].

During the Coronavirus epidemic, parents have become the main source of support for prevention, intervention and schools are no longer as involved in the prevention and intervention. Due to remote learning and COVID-19 modifications it is not possible to provide support in the traditional didactic way. Aside from this, prevention methods are not adequate to address this rising issue as well as the changes due to COVID-19. Past prevention and intervention methods have focused on helping the individual instead of addressing the system. However, changes due to COVID and rising suicide rates have created a crisis to reconsider the issue and more effective measures to address this public health concern. Now, instead of looking at just the individual as the 'problem', the evidence we have demonstrates the significance of addressing family relationships, intimate partner relationships, and crises as well as other system factors. Instead of viewing suicide and suicidal ideation as symptomatic of only an individual issue, it is important to view it as a systemic concern. One that considers the role of the family, close relationships and connections, job and life factors in contributing to suicidal ideation [28].

Approaches such as Attachment-Based Family Therapy view the family as the solution to the issue of suicidal ideation, and not the individual as the problem. COVID-19 has created pressure and changes to the rates of psychological distress and suicidal ideation and thereby has helped reveal how the issue is symptomatic of a systemic concern rather than only an individual issue. Another important concern aside from suicide attempts is suicidal ideation. A newer approach that has proved effective for addressing suicidal ideation and suicide attempts is the Collaborative Assessment and Management of Suicidality (CAMS). This is a treatment framework that provides risk assessment, stabilization planning and treatment of patient defined suicidal drivers and has proven effective across a range of suicidal states. Some critical aspects of CAMS are building a strong therapeutic alliance to create patient-motivation and address patient-defined suicidal drivers. Some paradoxical data on suicide rates in the United States are that 80% of suicides for individuals age 65 and older are completed by men, and are highest for older adult men of European decent. Overall, suicide is most common among older Caucasian males. This is a paradox as they are also the most socially and economically privileged, with the least severe aging adversities [29].

Currently with Coronavirus, in the U.S. suicide rates are at their highest levels since World War II. The factors that contribute to the exceptionally high rates of suicide include social isolation, reduced access to religious services, economic stress and general national anxiety and disease largely connected to a toxic political climate and social disruption and instability [27]. With Coronavirus there has also been a rise of healthcare provider suicide rates, firearm sales, and suicidal ideation for individuals that have contracted COVID-19, individuals with preexisting conditions/immune-compromised and individuals with substance abuse histories. With Coronavirus there has been an increase in the number of youth with suicidal ideation as well as ethnic minorities. Specifically, there has been a significant rise in suicidal ideation among African American children during Coronavirus [30].

Buffers in the rise of suicide rates include having immediate support, social support, planning for the future, core values/beliefs, connection and engagement with a healthcare provider to provide support, and a sense of purpose. For individuals who demonstrate suicidal ideation, it is important to be present and engaged, stay calm and not react, validate an individual's experiences, evaluate any intent, feelings of hopelessness, burden and disconnection. Some suggestions for how to manage your emotions during COVID-19 are to pay attention to your body and notice what you need. This includes things such as eating well, eating healthy foods, staying properly hydrated, and being creative about daily exercise. Another suggestion is to avoid catastrophizing unnecessarily. Our heightened exposure to media and news outlets with negative news to share has increased the amount of negativity we are being individually and collectively exposed to. This increase in negative and disturbing media due to COVID-19 and COVID-19 related events can be worsened by binging on news from television, radio, social media, emails and more [31].

Other suggestions for managing emotions, preventing suicidal ideation and psychological distress include managing a balance of alone time and social time. This alone time can include reading, journaling, meditating or other activities you may enjoy doing independently. Social time during COVID-19 restrictions may include connecting with individuals through telephone, FaceTime, Skype, Facebook, Zoom, and other video or social networking platforms. It can also include time spent face to-face with others you live with and family that have also been quarantined. Additional suggestions include being aware and mindful of your emotions. Mindfulness and awareness of your emotions is important for noticing and early intervention of how you are feeling, what you are feeling and where you are feeling it. Very often there is a misunderstood belief that it is necessary to react to what we are thinking and feeling and do something in reaction to these emotions. Yet with mindfulness it is sufficient to just be aware of our emotions and notice them. Just noticing what we are thinking, feeling and experiencing is sufficient for reducing negative emotions and distress and increasing positive emotions. Lastly, it is also important to have a plan in place that has been thought out ahead of time. Similar to a safety plan, but for your wellbeing and emotions, this plan addresses coping mechanisms, how to come back to a place of calm and emotional stability after you are feeling triggered by news and current events, or by the stress of COVID-19 and all the restrictions put in place as a result of it [31].

If you find you are still struggling with psychological distress teletherapy is a viable addition for managing distress. In a study conducted by Baumel et al. [32], cognitive rehabilitation and cognitive-behavioral therapy through the use of smartphone apps was shown to be beneficial in reducing psychological distress [32]. Also, a study conducted in the UK, found that teletherapy was as effective in reducing PTSD symptoms as in-person interventions for veterans [33]. A study led by Sproch and Anderson [34] found that evidence-based teletherapy interventions led by a teletherapy-trained therapist led to significant improvements of eating disorder symptoms [34]. Teletherapy and remote support can also be improved by reassuring individuals who are suffering that any technological concerns that arise will be addressed immediately and to provide options ahead of time for back-up remote support. However, during this time with COVID-19 if an individual expresses emotional distress to you, do not provide superficial reassurance, or ignore or avoid discussions on strong feelings. Even if they feel overwhelming, it is important to address strong feelings directly. Also, try to not be reactive and notice your own anxiety or stress in relation to how the individual is feeling. If you are uncomfortable hearing about emotional disturbances or suicidal ideation and become reactive as a result of this, it will not be supportive and can worsen the situation and emotional distress. Specifically, it is important to acknowledge and validate someone's distress, and discuss a plan for how to best provide support. It is not necessary to act like a mental health professional. Being present, staying calm and assisting in an emotional safety plan is something that we can all do when confronted with these challenges by someone [31].

Certain disaster mental health models are effective for these secondary and tertiary mental health effects related to COVID. Taking a trauma-informed approach is important for any work that is done. Trauma-informed care is care that "is grounded in and directed by a thorough understanding of the neurological, biological, psychological, and social effects of trauma and violence on humans and is informed by knowledge of the prevalence of these experiences in persons who receive mental health services" [35]. A trauma informed approach can be applied in traditional healthcare settings as well as behavioral health and social work environments. A seven-stage model designed by Dr. Albert Roberts is comprised of a biopsychosocial and lethality assessment, establishment of rapport, identification of problem dimensions, dealing with feelings, exploring alternative coping mechanisms, creating an action plan and a follow-up plan. Key components of the 7-stage crisis intervention model include validating the feelings of the client, establishing a rapport, understand additional factors that may be impacting the client such as biological, psychological and social concerns. Also, find out what the crisis is and create an action plan, work to understand what they are willing and able to do to confront the crisis they are in, the resources they have available to support them in their crisis, establish a referral network to address the crisis they are in and provide support to help with confidence building. In addition to this, there is an effort to actually implement the plan and follow-up everything they have done and everything they have been through. The aim of this model is to provide a quick but effective intervention for individuals in crisis situations. These crisis situations can be a matter of perception and can include issues such as loss of job, loss of relationship, significant life changes. Due to overwhelm in the mental and physical health system, crisis intervention methods that are rapid are needed to quickly address issues [36].

Other crisis interventions include cognitive-behavioral exercises to help individuals alter negative perspectives and reframe perception of situations into a more positive light. Mindfulness-Based Cognitive Therapy for example can be helpful for addressing depression, anxiety, unhealthy cyclical patterns, and mindfully training yourself to easily and quickly develop new or different perspectives to negative situations. Cognitive Behavior Therapy, Recovery Oriented Cognitive Behavior Therapy, and Rational Emotive Behavior Therapy are multiple kinds of cognitive behavioral approaches that work to reframe and positively direct thought processes. This can be very important for anxiety, depression or other issues arising from negative thought processes and stress especially during COVID. Cognitive Behavior Therapy works to reframe thought patterns that may lead to anxiety and depression. Recovery Oriented Cognitive Behavior Therapy focuses on inserting positive new thoughts and beliefs into thought patterns. It is focused on adding more positive self-concepts into everyday thinking such as:

I am good

I am valuable

I can handle anything that happens in my life

I am resilient,

I matter and my life is important

…etc.

So instead of the focus being identifying negative thought and belief patterns and working to reframe them, Recovery-Oriented Cognitive Behavioral Therapy goes directly to a focus on positive thoughts and adapting to inserting repeated positive thoughts. Rational Emotive Behavior Therapy works to identify negative beliefs and core belief systems and find other alternatives to these beliefs so that the individual can consider other perspectives and question assumed belief systems for how the world works. All of these methodologies work to interrupt assumed perspectives and ideas about life and your world so that you can see through a more positive perspective and reduce anxiety, depression, and other destructive difficult emotions [37].

Eye movement desensitization and reprocessing (EMDR) is a rapid therapeutic intervention to address trauma, distress and negative emotions. EMDR is a trauma-informed approach to assist individuals that are suffering from psychological distress rapidly, and is effective for those who do not desire or want to talk about what is upsetting them. It is a rapid, non-verbal approach to address psychological distress and stress. This approach is also used in international disaster related settings to help individuals that have suffered trauma, disasters, and tremendous stress. It has proven effective for brief interventions or situations where mental health providers can only come and provide services on a short-term basis and individuals are needing rapid results [38].

Another effective intervention for managing severe stress, psychological distress or other issues related to COVID is short-term psychopharmacological treatment. Anti-depressants or anti-anxiety medication are effective for reducing anxiety, depression and stress that can contribute to the creation of worse conditions such as panic attacks, Post-Traumatic Stress Disorder, eating disorders, suicidal ideation or suicidal attempt, clinical depression, relationship and intimate partner concerns, sleep issues and more. By addressing anxiety and depression through psychopharmacological treatment, it is possible to reduce the chances of symptoms worsening or building to something more destructive and harder to treat [38].

Other alternative approaches include hypnotherapy and on-going mindfulness meditation techniques. Hypnotherapy can be an effective way to address subconscious beliefs and thoughts, as well as reframe and alter negative beliefs to positive beliefs. Hypnotherapy can include certain aspects of EMDR and cognitive behavior exercises with safe place visualizations, insertion of positive belief systems and healing from trauma and distress. Mindfulness meditation can be effective for addressing racing thoughts as well as disturbing and upsetting thought patterns. Both hypnotherapy and mindfulness meditation effectively reduce the fight or flight response from the amygdala and calm the central nervous system. This is effective for improving resilience when confronted with increased levels of stress. It also helps to bring clarity to unexpected challenges and overwhelming emotional memories or occurrences. Mindfulness meditation provides individuals with tools they can use any time they are experiencing stress and emotional difficulty [38, 39].

3.4.2 Psychological First Aid

Another mental health framework that can be effective for addressing psychological distress related to the Coronavirus and issues connected to the Coronavirus include Psychological First Aid. Psychological First Aid is an evidence-informed model used for disaster response to assist individuals in the initial hours and days after an emergency, disaster or act of terrorism [40]. Psychological First Aid (PFA) is an evidence-informed method to help children, adolescents, adults and families after there is a disaster or mass trauma such as terrorism, violence due to protest or a pandemic (for example). It is based on an understanding that individuals can and often are negatively impacted by traumatic events whether as a survivor, witness, responder, or in some way connected to the event as a family member, loved one or partner of a survivor. PFA was created by the National Child Traumatic Stress Network and the National Center for PTSD. The PFA approach does not take the perspective that all individuals who experience a trauma will develop severe mental health issues or long-term issues. Rather, it takes the approach that survivors and those connected to them can experience a variety of emotions and reactions. The reactions include physical, psychological, behavioral and spiritual. Negative reactions that are severe enough can reduce the ability to cope in healthy ways [40].

Core principles of PFA include assistance and intervention in the days or weeks following a traumatic event. With the PFA approach, providers need to take a flexible, adaptive approach. The amount of time a provider spends with a client in the days and weeks following a disaster or traumatic event should be dependent on the survivors needs and issues. Psychological First Aid is meant to be applied in a variety of settings and circumstances. These situations may include general population shelters, special needs shelters, field hospitals, medical emergency departments, crisis hotlines and phone banks, feeding locations, and other places for emergency assistance [40].

Core actions of PFA include contact and engagement where the provider initiates contact to a survivor or responds to contact from a survivor. Another core action of PFA is safety and comfort. The goals of PFA are to increase immediate and longer-term safety and give physical and emotional comfort. A third component is to provide stabilization to survivors and help calm distressed survivors. A fourth component is obtaining information on immediate needs and concerns of survivors. The fifth aspect is offering practical help to survivors to assist with immediate concerns and needs. The sixth aspect is to help provide connection to social supports for the survivor such as family members, friends and community helping resources. The seventh component is for the provider to give information on coping to the survivor such as what someone can expect with stress and trauma reactions and how to reduce distress and strengthen healthy coping mechanisms. The eighth and final core component of PFA is helping to link survivors to services that are needed at present or in the future [40].

3.4.3 Culture-Centered Disaster Mental Health Counseling

Culture-Centered Disaster Mental Health Counseling is a potentially effective approach to consider especially concerning the recent violence, protests and cultural concerns being experienced particularly in the United States. In this framework there are seven steps to consider. The first step in this model is awareness, and having participants recognize that they bring their own biases into the environment and into their experience and perspective of the world. The second aspect of this model is respect. In the second step participants recognize that community members have equally valid realities and funds of knowledge. It provides respect and acknowledgement of the inherent humanity and value of each and every individual. The third component of this model is context. When considering context, individuals recognize the sociopolitical context that emotions, experiences and psychological distress occur. Understanding the sociopolitical context in which emotions occur is helpful for a fuller comprehension of what is impacting an individual. The fourth step is integration. In this part of the model, individuals integrate knowledge they have of sociopolitical context and respect into their clinical experience. The fifth component of the model is empowerment. With empowerment, individuals are able to concentrate and act on an objective of empowerment in their perspective and life. In the sixth step, individuals create a plan of action. The seventh step is transformation where individuals integrate their insights and experiences into their personal and professional identities. In this model, counselors reach out to culturally diverse individuals and communities and better understand and intervene with culturally diverse clients. This model promotes awareness, knowledge and skills to address client needs through a culturally sensitive lens [41].

3.4.4 Anxiety

There is an increase in anxiety and stress related anxiety both in the United States and worldwide due to COVID. Anxiety is rising for a variety of reasons. Much of this includes feelings of fear and uncertainty about the future and feeling that not only is the present moment uncertain but the future is uncertain for an unknown amount of time. There is anxiety related to attempting to engage in social interactions and fear about being infected with the virus or others having COVID-19 and being asymptomatic. There can also just be a general anxiety from adults and children about the virus in general. Public policies and public health actions can make individuals, families and couples feel isolated and lonely. Social connections are important as a buffer for stress and managing feelings of uncertainty. Social isolation can increase feelings of stress and anxiety. These emotions can also create challenges with sleeping and eating, issues with chronic physical health and mental health conditions, increase use of tobacco, alcohol and other substances. There can also be anxiety increase and worry about loved ones, financial situation and jobs, and loss of support services that are relied on [42].

In the United States here are some organizations that can help support individuals suffering from stress and anxiety or other psychological issues. The National Suicide Prevention Lifeline at 1-800-273-TALK (8255) for English, 1-888-628-9454 for Spanish. Disaster Distress Helpline at 1-800-985-5990 (press 2 for Spanish), or text TalkWithUs for English or Hablanos for Spanish to 66,746. National Domestic Violence Hotline 1-800-799-7233 or text LOVEIS to 22,522. National Child Abuse Hotline 1-800-4AChild (1-800-422-4453) or text 1-800-422-4453. Two emergency numbers are 911 for general emergencies and crises or 988 for mental health crises. The National Child Abuse Hotline can be reached at 1-800-4AChild (1-800-422-4453) or text 1-800-422-4453. The National Sexual Assault Hotline can be reached at 1-800-656-HOPE. The Eldercare Locator can be contacted at 1-800-677-1116. The Veteran's Crisis Line can be reached at 1-800-273-8255. In order to find a health care provider or treatment for substance use and mental health concerns, SAMSHSA National Helpline at 1-800-662-HELP can be contacted.

3.4.5 Depression

In a very recent study, published in September, 2020, an assessment of 1441 respondents was evaluated to investigate the mental health impacts of the COVID-19 pandemic. The investigation included a survey study of 5065 respondents before the pandemic and 1441 respondents during the pandemic to assess a comparison for statistical significances. This investigation revealed that depression rates are three times higher during the pandemic than before. Particular factors such as a lower income, less than $5000 in savings, and exposure to more stressors was associated with higher levels of depression. The outcomes from this investigation suggest that more services should be provided and created for mental illness treatment with a particular focus on vulnerable populations [24].

3.4.6 Addiction

There is an increase in addiction and potential for on-going increases in addiction rates. Physical distancing and isolation are key issues with the increase of addiction rates. It causes mental health issues and difficulties with coping for even individuals who did not previously struggle with mental illness. Another part of the rising issues are the difficulties that children are experiencing with uncertainty, changes in their routine and social isolation. COVID-19 led to school cancellations for 30 million children and has led to food insecurity for many children that rely on breakfast and/or lunch meals provided at school. For other children living with household abuse, being quarantined is keeping children stuck with abusers. It can also mean that children are experiencing important events such as graduations [43].

Certain protective factors for teen addiction include parental monitoring, academic success and neighborhood connections. However, protective factors have been reduced with COVID-19. In addition to this the impact of the pandemic is increasing the rate of childhood trauma and this also increases addiction risk. Childhood trauma is associated with future substance use disorders and mental illness. It also increases the exposure to potentially traumatic experiences such as harassment, abuse, illness, and poverty. The healthcare system is currently under strain and individuals with addiction are struggling with serious consequences due to lack or limited resources. In addition to this, not only is there a strain in our healthcare system for addiction resources, there are also greater risks for individuals who smoke tobacco, marijuana, vape, use opioids or methamphetamines with COVID-19. Individuals who use these substances are at greater risk of respiratory or pulmonary issues with the pandemic of COVID-19. Individuals with substance use disorder (SUD) are particularly at risk for infection for the COVID-19 virus and associated health challenges and also at higher risk for not obtaining the proper care due to prejudice and limited resources for SUD individuals. Vaping can cause lung illnesses such as popcorn lung, and e-cigarette aerosols can damage lung tissue, cause inflammation, and decrease the lung's ability to fight infection. Aside from these risks, chronic respiratory disease increases the possibility of fatal overdose for individuals that use prescription opioids. Methamphetamine causes pulmonary damage, pulmonary hypertension and cardiomyopathy [43].

There is more and more demand for rehab centers. We are seeing a huge spike in Opioid addiction in the United States and its directly from COVID and all the stress, unemployment, economic issues and uncertainty so many are facing. People that have not had addiction issues in the past have started to drink more, use more prescription drugs like benzos or significantly increase use of drugs like marijuana.

Social isolation and being stuck at home or having stress from family or relationships is part of more people having addiction. Also, with more free time being at home a lot of people are starting to realize they have a problem when maybe they didn't realize they had one before…whatever all the reasons are we can't keep up with the demand to create new rehabs. We just started one and three more at least to go. We estimate we need at least 150 new rehabilitation facility beds in the area to meet demand in the next year, but it could end up being more.

—Anonymous statement from an executive of a rehabilitation facility company on the state of COVID and addiction

3.5 Systems, Complexity and COVID

The impacts of such events as COVID cannot be seen in isolation. WEF Global Risks reports have clearly articulated the hyperconnected risks that permeate the world and reflect the inherent '…fragility and vulnerabilities that lie within the social/technological/economic/political/ecological interdependent systems' [44]. It is through these underlying networks that Helbing ([45]: 51) argues that we have '…created pathways along which dangerous and damaging events can spread rapidly and globally' and thereby has increased systemic risks.

A Chatham House report, 'Preparing for High Impact, Low Probability Events', found that governments and businesses remain unprepared for such events [46]. Helbing ([45]: 51) argues that:

Many disasters in anthropogenic systems should not be seen as 'bad luck', but as the results of inappropriate interactions and institutional settings. Even worse, they are often the consequences of a wrong understanding due to the counter-intuitive nature of the underlying system behaviour.

As described in Ramos et al. ([8]: 3), 'Systems thinking offers a more-integrated perspective and a number of proven concepts, tools, and methods to improve our understanding of potentially threatening complex, systemic issues. Systems thinking can improve the prospects for successful policy outcomes by offering a methodology and a range of simple tools to disaggregate, understand, and act on connected systemic issues, while taking proper account of the critical linkages between them. This can enable us to understand better the behaviour of complex, dynamic systems so as to anticipate their evolution, assess and manage risks, and decide how and where to intervene through targeted policies'.

Systems thinking brings to the forefront an understanding of interdependencies, interconnectivity and the identification of leverage points to manage a crisis. Figure 1 presents a systems map of some of the reported secondary and tertiary effects of COVID-19.

Of note are the relationships between societal fragility and societal impacts. Addressing issues to support mitigation, preparedness, response and recovery strategies requires a system mapping to unearth potential unintended consequences and to test for leverage points in the system. In a system, you cannot do just one thing, the effects resonate across the system nodes.

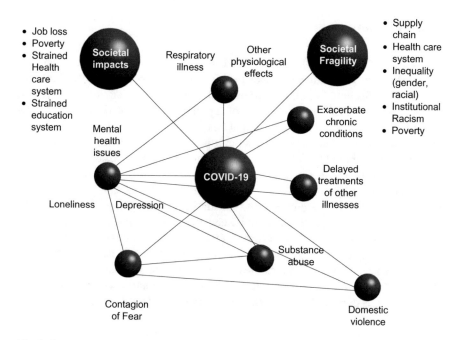

Fig. 1 Systems map of first, second order and tertiary effects of COVID-19

3.6 A Framework that Addresses Direct and Indirect Effects

The more research that continues in understanding the direct and indirect effects of COVID-19 the more evident it becomes of the need to create a framework that incorporates these considerations into the current healthcare model and intervention methods moving forward. A report titled *"A Framework for Educating Health Professionals to Address the Social Determinants of Health"* published by the National Academies Press points to the importance of incorporating additional social determinants of health into the public health guidelines and framework. Specifically, the National Academies recommended education and training related to social isolation and loneliness for healthcare workers. Also, to include the concerns of social isolation and loneliness in the U.S. Department of Health and Human Services key health interventions.

The World Health Organization has a framework known as the Health in All Policies and many other health providers have adopted a similar approach. This approach integrates health policies from multiple sectors and areas. The Health in All Policies methodology can include collaboration across disparate sectors such as transportation, energy, urban planning, housing, waste management, industry, health sector, international, local municipalities, NGOs, donors and civil society.

4 Conclusion

The World Health Organization has declared the rapid spread of COVID19 around the world a global public health emergency [47]. This certainly connects with Weick and Sutcliffe [3] reflection: '…unexpected events often audit our resilience. Everything that was left unprepared becomes a complex problem'. The Covid-19 outbreak has resulted in considerable impact on global health security, human security and has also impacted the global economy. As described in OECD ([2]: 2) 'the pandemic has reminded us bluntly of the fragility of some of our most basic human-made systems. … we have witnessed the cascading collapse of entire production, financial, and transportation systems, due to a vicious combination of supply and demand shocks'. This chapter has examined the secondary and tertiary impact of COVID-19 on global health security. Key is the requirement for a systems thinking lens to support sensemaking in the formulation of intervention strategies for crisis management. It also points to the requirement for reflection on the inherent vulnerabilities that have been revealed when COVID-19 stress tested our national and global societal systems. We are reminded of the words of Taleb [4] **'not seeing a tsunami or an economic event coming is excusable; building something fragile to them is not'**.

References

1. Felbab-Brown V, Caulkins JP, Graham C, Humphreys K, Pacula RL, Pardo B, Reuter P, Stein BD, Wise PH (2020) The opioid crisis in America. Domestic and international dimensions. Brookings paper series (June 2020)
2. OECD (2020) A systemic resilience approach to dealing with Covid-19 and future shocks. New Approaches to Economic Challenges (NAEC), 28 April 2020
3. Weick KE, Sutcliffe KM (2007) Managing the unexpected: resilient performance in an age of uncertainty, 2nd edn. Wiley, San Francisco
4. Taleb NT (2014) Antifragile: things that gain from disorder. Random House, New York
5. Massaro E, Ganin A, Perra N, Linkov I, Vespignani A (2018) Resilience management during large-scale epidemic outbreaks. Nat Sci Rep 8:1859
6. Masys AJ (ed) (2016) Disaster forensics: understanding root cause and complex causality. Springer Publishing
7. Jackson MC (2003) Systems thinking: creative holism for managers. Wiley, West Sussex
8. Ramos G, Hynes W, Müller J-M, Lees M (ed) (2019) Systemic thinking for policy making—the potential of systems analysis for addressing global policy challenges in the 21st century. OECD SG/NAEC(2019)4
9. WEF (2020) COVID-19 risks outlook: a preliminary mapping and its implications. https://www3.weforum.org/docs/WEF_COVID_19_Risks_Outlook_Special_Edition_Pages.pdf
10. Weick KE (1995) Sensemaking in organizations. Sage, Thousand Oaks, CA
11. Ancona DL (2011) SENSEMAKING *framing and acting in the unknown*. In: Handbook of teaching leadership. Sage Publications, Inc., Thousand Oaks, CA, pp 3–20
12. Cao X (2020) COVID-19: immunopathology and its implications for therapy. Nat Rev Immunol 20(5):269–270
13. Cain DW, Cidlowski JA (2020) After 62 years of regulating immunity, dexamethasone meets COVID-19. Nat Rev Immunol 1–2

14. Wu YC, Chen CS, Chan YJ (2020) The outbreak of COVID-19: an overview. J Chin Med Assoc 83(3):217
15. Quinn A, Laws M (2020) Addressing community needs and preparing for the secondary impacts of Covid-19. NEJM Catal Innov Care Delivery
16. Wenham C, Smith J, Morgan R (2020) COVID-19: the gendered impacts of the outbreak. The Lancet 395(10227):846–848
17. Taub A (2020) A new Covid-19 crisis: domestic abuse rises worldwide. The New York Times 6
18. Bradbury-Jones C, Isham L (2020) The pandemic paradox: the consequences of COVID 19 on domestic violence. J Clin Nurs
19. Kofman YB, Garfin DR (2020) Home is not always a haven: the domestic violence crisis amid the COVID-19 pandemic. Psychol Trauma: Theor Res Pract Policy
20. Campbell AM (2020) An increasing risk of family violence during the Covid-19 pandemic: strengthening community collaborations to save lives. Forensic Sci Int Rep 100089
21. Banerjee D, Rai M (2020) Social isolation in Covid-19: the impact of loneliness
22. Hwang TJ, Rabheru K, Peisah C, Reichman W, Ikeda M (2020) Loneliness and social isolation during the COVID-19 pandemic. Int Psychogeriatr 1–15
23. Killgore WD, Cloonen SA, Taylor EC, Dailey NS (2020) Loneliness: a signature mental health concern in the era of COVID-19. Psychiatry Res 113117.
24. Ettman CK, Abdalla SM, Cohen GH, Sampson L, Vivier PM, Galea S (2020) Prevalence of depression symptoms in US adults before and during the COVID-19 pandemic. JAMA Netw Open 3(9):e2019686. https://doi.org/10.1001/jamanetworkopen.2020.19686
25. Chen S, Bonanno GA (2020) Psychological adjustment during the global outbreak of COVID-19: a resilience perspective. Psychol Trauma: Theor Res Pract Policy 12(S1):S51
26. McGinty EE, Presskreischer R, Han H, Barry CL (2020) Psychological distress and loneliness reported by US adults in 2018 and April 2020. JAMA
27. Czeisler MÉ, Lane RI, Petrosky E, Wiley JF, Christensen A, Njai R et al (2020) Mental health, substance use, and suicidal ideation during the COVID-19 pandemic—United States, June 24–30, 2020. Morb Mortal Wkly Rep 69(32):1049
28. Kopelovich SL, Monroe-DeVita M, Buck BE, Brenner C, Moser L, Jarskog LF et al (2020) Community mental health care delivery during the COVID 19 pandemic: practical strategies for improving care for people with serious mental illness. Community Mental Health J 1–11
29. Jobes DA, Joiner TE (2019) Reflections on suicidal ideation
30. Reger M, Stanley I, Joiner T (2020) Suicide mortality and coronavirus disease 2019—a perfect storm? JAMA Psychiatry 77. https://doi.org/10.1001/jamapsychiatry.2020.1060
31. Kleiman EM, Liu RT (2013) Social support as a protective factor in suicide: findings from two nationally representative samples. J Affect Disord 150(2):540–545
32. Baumel A, Correll CU, Hauser M, Brunette M, Rotondi A, Ben-Zeev D, Gottlieb JD, Mueser KT, Achtyes ED, Schooler NR, Robinson DG, Gingerich S, Marcy P, Meyer-Kalos P, Kane JM (2016) Health technology intervention after hospitalization for schizophrenia: service utilization and user satisfaction. Psychiatr Serv 67(9):1035–1038. https://doi.org/10.1176/appi.ps. 201500317. Epub 2016 Jun 1. PMID: 27247171 (2016 Sep 1)
33. Turgoose D, Ashwick R, Murphy D (2017) Systematic review of lessons learned from delivering tele-therapy to veterans with post-traumatic stress disorder. J Telemed Telecare 24(9):575–585. https://doi.org/10.1177/1357633X17730443. Epub 2017 Sep 29. PMID: 28958211 (2018 Oct)
34. Sproch LE, Anderson KP (2019) Clinician-delivered teletherapy for eating disorders. Psychiatric Clin 42(2):243–252
35. Roberts AR, Yeager KR (2007) Crisis intervention with victims of violence. In: Roberts AR, David W (eds) (2007) From social work in juvenile and criminal justice settings, pp 95–105. Springer, See NCJ-217866
36. Roberts AR, Ottens AJ (2005) The seven-stage crisis intervention model: a road map to goal attainment, problem solving, and crisis resolution. Brief Treat Crisis Interv 5(4):329
37. Stanley B, Brown G, Brent DA, Wells K, Poling K, Curry J et al (2009) Cognitive-behavioral therapy for suicide prevention (CBT-SP): treatment model, feasibility, and acceptability. J Am Acad Child Adolesc Psychiatry 48(10):1005–1013

38. Gauhar YWM (2016) The efficacy of EMDR in the treatment of depression. J EMDR Pract Res 10(2):59–69
39. Luoma JB, Villatte JL (2012) Mindfulness in the treatment of suicidal individuals. Cogn Behav Pract 19(2):265–276
40. Uhernik JA, Husson MA (2009) Psychological first aid: an evidence informed approach for acute disaster behavioral health response. Compelling Counse Interventions: VISTAS 200(9):271–280
41. West-Olatunji C, Yoon E (2013) Culture-centered perspectives on disaster and crisis counseling. J Asia Pacific Counsel 3(1)
42. Lee SA, Mathis AA, Jobe MC, Pappalardo EA (2020) Clinically significant fear and anxiety of COVID-19: a psychometric examination of the Coronavirus anxiety scale. Psychiatry Res 113112
43. Volkow ND (2020) Collision of the COVID-19 and addiction epidemics Washington (DC): National Academies Press (US); 2016 Oct 14. ISBN-13:978-0-309-392624 ISBN-10:0-309-39262-4
44. Masys AJ, Ray-Bennett N, Shiroshita H, Jackson P (2014) High impact/low frequency extreme events: enabling reflection and resilience in a hyper-connected world. In: 4th international conference on building resilience, 8–11 Sept 2014, Salford Quays, United Kingdom (Procedia Econ Finance 18:772–779)
45. Helbing D (2013) Globally networked risks and how to respond. Nature 497:51–59
46. Lee B, Preston F, Green G (2012) Preparing for high-impact, low-probability events lessons from Eyjafjallajökull—A Chatham House Report. https://www.chathamhouse.org/sites/def ault/files/public/Research/Energy,%20Environment%20and%20Development/r0112_highim pact.pdf
47. Dryhurst S, Schneider CR, Kerr J, Freeman ALJ, Recchia G, van der Bles AM, Spiegelhalter D, van der Linden S (2020) Risk perceptions of COVID-19 around the world. J Risk Res 23(7–8):994–1006
48. West-Olatunji C, Henesy R, Varney M (2014) Group work during international disaster outreach projects: a model to advance cultural competence. J Spec Group Work 40(1):38–54
49. Zhang J, Lu H, Zeng H, Zhang S, Du Q, Jiang T, Du B (2020) The differential psychological distress of populations affected by the COVID-19 pandemic. Brain, Behav Immun

Sensemaking and Disaster Forensics: An Examination of Cholera Epidemics

Daniel Hutchinson, Jeegan Parikh, and Anthony J. Masys

Abstract Cholera remains a public health issue in regions around the world. 'The epidemiology of cholera in the areas of Asia, Africa and the Americas where the disease occurs, continues to evolve' (Deen et al. in Vaccine 38:A31–A40, 2020 [1]). In managing public health crisis stemming from Cholera outbreaks, epidemiological data are crucial for decision-making pertaining to tactical and strategic interventions. Sensemaking as a process is key in surveillance, modeling, weak signal detection, mapping of high-incidence areas and design of intervention strategies across the spectrum of disaster management (mitigation, preparedness, response and recovery). Weick (Weick in Sensemaking in organizations, Sage, Thousand Oaks, CA, 1995 [2]) refers to sensemaking in terms of '…how we structure the unknown so as to be able to act in it. Sensemaking involves coming up with a plausible understanding—a map— of a shifting world; testing this map with others through data collection, action, and conversation; and then refining, or abandoning, the map depending on how credible it is' (Ancona in Handbook of teaching leadership. Sage Publications Inc, Thousand Oaks, CA, pp 3–20, 2011 [3]). Public health disaster events require leaders and planners to engage in making sense of the impact and inherent vulnerabilities. As articulated in (Nowling et al. in J Appl Commun Res, 2020 [4], 'they scan their environments for cues of changes and operational threats. After identifying the threat or change, they communicate and work collaboratively to determine the most plausible meaning of these cues (Weick in Admin Soc 38(4):628–652, 1993 [5])'. This chapter explores Cholera epidemics and the critical role sensemaking places in both tactical and strategic interventions. Systems thinking plays a key role in the sensemaking process and approach and is central to the discussion.

Keywords Cholera · Sensemaking · Systems thinking · Disaster forensics · Disaster management

D. Hutchinson · J. Parikh · A. J. Masys (✉)
College of Public Health, University of South Florida, Tampa, FL, USA
e-mail: Anthony.masys@gmail.com

A. J. Masys
International Centre for Policing and Security, University of South Wales, Pontypridd, UK

A. J. Masys (ed.), *Sensemaking for Security*, Advanced Sciences and Technologies for Security Applications, https://doi.org/10.1007/978-3-030-71998-2_5

71

1 Introduction

Cholera remains both a recurrent and emergent public health issue in regions around the world. As described in Deen et al. [1], 'Seven pandemics of cholera have been recorded since the first pandemic in 1817, the last of which is ongoing. Overcrowding, poverty, insufficient water and sanitation facilities increase the risk for cholera outbreaks'. In managing public health crisis stemming from Cholera outbreaks, epidemiological data is crucial for decision-making pertaining to tactical and strategic interventions. Sensemaking as a process is key in surveillance, modeling, weak signal detection and mapping of high-incidence areas. Weick [2] refers to sensemaking in terms of '…how we structure the unknown so as to be able to act in it. Sensemaking involves coming up with a plausible understanding—a map—of a shifting world; testing this map with others through data collection, action, and conversation; and then refining, or abandoning, the map depending on how credible it is' [3].

Public health disaster events require leaders and planners to engage in making sense of the impact and inherent vulnerabilities. As articulated in Nowling et al. [4], 'they scan their environments for cues of changes and operational threats. After identifying the threat or change, they communicate and work collaboratively to determine the most plausible meaning of these cues [5].

2 Overview of Cholera

Cholera has been a major public health issue affecting the most vulnerable populations across the globe. For example, the current cholera epidemic in Yemen has been ongoing since 2016. According to the World Health Organization (WHO), the total number of suspected cholera cases reported from January 2018 to May 2020 were 1,371,819.

Cholera is a waterborne disease that is produced by *Vibrio Cholerae* bacteria and is naturally present in the environment. *V. Cholerae* bacteria generally proliferates in areas that are warm and moderately saline—these areas are typically on the coast such as estuaries (where fresh water meets seawater) [6]. It is also known that Cholera outbreaks may be initiated when man disrupts a cholera reservoir in the natural environment or in simpler terms drinks contaminated water. From here on out, transmission from person to person occurs. The WHO states that:

> The faeces, and often the vomitus, of cholera patients contain high concentrations of cholera bacteria, which can then contaminate water and food sources when passed back into the environment, with the potential for causing cholera outbreaks… that passage through the human host transiently increases the infective potential of *V. cholerae* by creating a hyper infectious state that is maintained soon after shedding, and which may contribute to the epidemic spread of the disease. [7]

Nevertheless, "direct transmission" between individuals is usually not the method that contributes to the spreading of cholera and does not significantly contribute to the burden of disease; instead, the faeco-oral transmission route is what needs urgent attention [7]. The statistics that support this are alarming, according to the WHO, at least 2 billion people across the globe utilize water that is contaminated with fecal matter [8]. Most of these individuals are found in parts of sub-Saharan Africa and Asia—areas where extreme poverty can also be found [7].

3 Sensemaking

Within the context of cholera outbreaks, sensemaking as a process is key in surveillance, modeling, weak signal detection, mapping of high-incidence areas and design of intervention strategies across the spectrum of disaster management (mitigation, preparedness, response and recovery). Weick [2] refers to sensemaking in terms of '…to how we structure the unknown so as to be able to act in it. Sensemaking involves coming up with a plausible understanding—a map—of a shifting world; testing this map with others through data collection, action, and conversation; and then refining, or abandoning, the map depending on how credible it is' [3].

Our mental models, (beliefs and assumptions about how the world works) shape what we see and how we label, categorize and make meaning thereby shaping our decisions and actions. Within the context of disaster and crisis management, sensemaking is about interpretation and action. It is a '… process through which individuals in an organization come to understand problems or events that are ambiguous, unexpected, new or confusing' [9].

Given the complex aetiology associated with Cholera, sensemaking must reveal the 'systemic risks' that are inherent in its emergence and persistence. With this in mind, systems thinking figures prominently as an approach to support sensemaking.

4 Systems Thinking

The reductionist paradigm has dominated most of classical science based upon a worldview of independence and linearity. Our linear mindset and reductionist approach to understanding complex problems fails. Jackson ([10]: 65) defines systems thinking paradigm as'…a discipline for seeing the 'structures' that underlie complex situations, and for discerning high from low leverage change…Ultimately, it simplifies life by helping us to see the deeper patterns lying beneath the events and the details'. In this complex problem landscape of public health [11] and the emergence and transmission of disease, systems thinking emerges as both a worldview and a process in the sense that it informs ones understanding regarding a system and can be used as an approach in problem solving ([12]: 5).

As described in Masys [13] 'Systems thinking' according to Senge [14] emphasizes interconnectedness, causal complexity and the relation of parts to the whole [15], thereby challenging traditional linear thinking and simple causal explanations.

Senge ([14]: 68) describes systems thinking as 'a discipline for seeing wholes…a framework for seeing interrelationships rather than things, for seeing patterns of change rather than static snapshots'. As a worldview, systems thinking recognizes that systems cannot be addressed through a reductionist approach that reduces the systems to their components. Systems thinking purports that, although events and objects may appear distinct and separate in space and time, they are all interconnected. As such, systems thinking has developed into a field of inquiry and practice [13].

Peters [16] argues that

> …at its core, systems thinking is an enterprise aimed at seeing how things are connected to each other within some notion of a whole entity. We often make connections when conducting and interpreting research, or in our professional practice when we make an intervention with an expectation of a result. Anytime we talk about how some event will turn out, whether the event is an epidemic, a war, or other social, biological, or physical process, we are invoking some mental model about how things fit together. However, rather than relying on implicit models, with hidden assumptions and no clear link to data, systems thinking deploys explicit models, with assumptions laid out that can be calibrated to data and repeated by others.

To put into context, a disease anywhere is a disease everywhere. Public health crises do not respect borders. As reported in Massaro et al. [17] 'The evaluation of vulnerabilities and consequences of epidemics is a highly dimensional complex problem that should consider societal issues such as infrastructures and services disruption, forgone output, inflated prices, crisis-induced fiscal deficits and poverty'.

5   Stress-Testing Communities: Revealing Vulnerabilities and Interdependencies

Disasters have the propensity to seriously paralyze the normal operations of a system. These systems being described are essentially the critical infrastructure and services that are needed to maintain normalcy and order with a country. Knap and Rusyn indicates that natural disasters bring the vulnerabilities of public health infrastructure to the surface.

> …public health outcomes range from exacerbation of chronic disease and stress, enhanced effect of infectious disease outbreaks particularly in developing countries, displaced communities and serious degradation of public health infrastructure. [18]

Developing countries in particular are most vulnerable to natural disasters due to the unsuspecting frailty of their critical infrastructure. As such, disasters such as earthquakes and tropical cyclones not only highlight all vulnerabilities but increase human misery and suffering [18].

The breakdown of societal infrastructure after major natural disasters such as hurricanes, floods and earthquakes or from man-made disasters (civil war, industrial accident) may increase the likelihood of a disease outbreak under conditions that stress the health care system such as an inadequate clean water supply, poor sanitation and neglected waste management. For example, after a massive earthquake, crush injuries are very common as people will become trapped under the debris. With extensive surgeries being done to treat the immediate effects of the disaster, a considerable strain will be placed on the attending medical facilities. The secondary and tertiary impacts of the disaster may emerge (cholera outbreak) thereby further stressing an already shocked healthcare system. Large populations of people may be displaced and may have to resort to available but polluted water sources. While a cholera outbreak does not have to occur after every disaster, the faster sanitation and water treatment infrastructure is restored, the less likely an outbreak will occur [19].

Tropical cyclones can result in serious damage to critical infrastructure and healthcare systems as do earthquakes. Severe tropical cyclones will place a strain on existing health care facilities and exacerbate weaknesses in the overall physical infrastructure particularly through the forces of strong wind and storm surge. Flooding during and after a tropical cyclone can also contribute to a cholera outbreak. This is due to contamination process of existing clean water supplies such as boreholes, wells and surface water storage. Similarly, to that of an earthquake; the breakdown of (and reluctance to repair) critical infrastructure after a tropical cyclone over a long period of time will increase the risk of a cholera epidemic.

Mozambique can be used as an example of how tropical cyclones can expose vulnerabilities of a country. During mid-March 2019, tropical cyclone Idai devastated south eastern Africa. Mozambique in particular was severely affected:

> Cyclone Idai [Category 2 storm] wiped out roads, bridges, and dams as it swept through Southeast Africa. The United Nations estimated that Cyclone Idai and subsequent flooding destroyed more than $773 million in buildings, infrastructure, and crops.

Cyclone Kenneth, (a category 3 storm) then ripped through Mozambique on April 25, 2019; damaging whatever infrastructure was left. Incidentally, Cyclone Kenneth was the strongest tropical cyclone to hit Africa [20]. Subsequently to the passage of these storms, a cholera outbreak occurred. Cambaza et al. indicates that the basic health, water and sanitation infrastructure was completely damaged by the second storm. The author also states that a cholera outbreak was expected by the government of Mozambique, due to the seasonal heavy rainfall that is associated with waterborne diseases [20]. The extenuating circumstances however (such as lack of clean water combined with thousands of internally displaced personnel) that manifested themselves after the disaster contributed to the outbreak. Herein lies the application of systems thinking as a key approach in understanding interdependencies and vulnerabilities in social infrastructure that can impact human security.

6 Outbreaks in African Countries

After analyzing countries such as Uganda, Zimbabwe and Sierra Leone, affected by cholera epidemics or with cholera endemicity, a few key features were observed.

These countries have been battling cholera while simultaneously dealing with other social and economic challenges which adds more complexity to the existing chaos.

Poverty is the most obvious feature amongst all these African countries. They also have relatively weak health care systems in addition to poor water and sanitation infrastructure. While the purpose of this chapter is not to go into a detailed comparison of said countries; it is important to mention these socioeconomic factors because of their contribution to their overall resilience. They must also be considered because they will impact whether these countries can follow through on the general cholera mitigation recommendations given by the WHO. For example, the conditions under which a cholera epidemic may be kindled are similar to that of several other diarrheal diseases. Therefore, a cholera epidemic or outbreak may be indicative of other disease threats of high public health importance such as shigellosis, typhoid and Hepatitis E. During the 2008–2009 cholera outbreak in Uganda, it was noted by Bwire et al. that districts within the capital city of Kampala that had the highest incidence of cholera, were also prone to these other 3 disease [21].

Another feature that was identified was that these countries were initially de-stabilized by civil war or political tension which disrupted the basic critical infrastructure that existed. After these disruptions, cholera and other diarrheal outbreaks began to occur. In Uganda for example, cholera cases were first observed when 757 cases were reported to the WHO in 1971 [21]. Incidentally, 1971 also happens to be the year when the Ugandan genocide began [22]. Today, Uganda continues to face cholera outbreaks.

Sierra Leone is another country that is currently dealing with the after-effects of a civil war. According to the publication, *Water Supply and Sanitation in Sierra Leone: Turning Finance into Services for 2015 and Beyond*, the World Bank indicates that, "A decade of civil war destroyed both economic and social infrastructure, creating competing demands on scarce national resources" [23]. While the country is recovering and trying to implement better water and sanitation programs, there are considerably large group of people that remain vulnerable due to lack of access to this basic human right—safe water.

Zimbabwe was slightly different, but the situation was more insidious with the political tension between two parties contributing to slow and steady decline in the water and sanitation infrastructure of the country. In short, ineffective governance was a root cause that led to several human rights violations. This eventually resulted in sewage being pumped into water mains leading to the contamination of potable water for residents of certain districts [24]. Under these conditions it was only a matter of time before the cholera outbreak exploded. Other dynamic pressures such as a failure in the healthcare delivery system combined with several other unsafe conditions (such as hyperinflation) contributed to furthering the cholera epidemic.

Additionally, these African countries have health care systems that are overburdened with other diseases such as HIV/AIDS and malaria. A cholera epidemic will put further pressure on the already burdened system and lead to weakening of the pre-existing infrastructure. Sierra Leone for example had to deal with one of the largest cholera outbreaks the country has experienced in its history in 2012. Subsequently in 2014, the Sierra Leone government had to deal with the Ebola outbreak while still having cholera cases. Dr. Seth Berkley, CEO of the GAVI Alliance commented on this issue in a WHO article: "Access to safe water and sanitation is limited, and the public health system, still recovering after the 2014 Ebola outbreak, is stretched" [7]. What emerges from this analysis is that cholera epidemics are weak signals pointing to systemic risks of poor and vulnerable public health systems and supporting critical infrastructure.

7 Case Studies (Region of Americas)

Before the 1991 cholera outbreak in Peru and its spread to other Latin American countries, there had been no previous outbreak in the regions of Americas for centuries [25, 26]. The 1991 Peru outbreak was followed by another outbreak in Haiti in 2010 resulting in two major cholera outbreaks in the region of Americas in three decades [25, 26]. The cholera outbreak in the Dominican Republic in 2010 and its associated Venezuelan outbreak in 2011 were the other minor limited outbreaks in the region [27]. In both the major outbreaks, there was controversy surrounding the start of the epidemic with the Chinese dumping of ballast being blamed in Peru while the United Nations (UN) peacekeeping force was held responsible for the introduction in Haiti [25, 28]. However, the main factors that led to the arrival, rapid transmission, and continuation of the cholera outbreak were fractures in critical public infrastructure including water, sanitation systems, and poverty [27, 25]. This makes the cholera outbreaks an opportunistic societal infection [27].

7.1 Haiti Outbreak (2010–2017)

On January 12, 2010, Haiti had a devastating earthquake that destroyed buildings, roads, water, and sanitation infrastructure including creating millions of internally displaced refugees [27]. This was followed by the sudden increase in patients with acute rice watery diarrhea and dehydration in October 2010 leading to the public announcement of the cholera epidemic on October 22, 2010 [28, 29]. The poor health indices of Haiti combined with low access to healthcare, safe water, and improved sanitation due to poverty contributed to the rapid spread of cholera associated with high initial case fatality rates [27]. Initial investigations led to the conclusion of many patients having consumed untreated contaminated water from nearby rivers [28]. The epidemiological studies by the US CDC and confirmation of *V. cholerae* serogroup O1 biotype Ogawa led to the source of water contamination being a UN peacekeeper from Nepal [28]. Although the factors of poverty, water, sanitation systems, and nearby rivers put Haiti at risk of cholera, there had been no outbreak of cholera since the 1800s because of a lack of presence of the agent in sufficient quantities [27, 30]. However, once the agent was introduced, the outbreak of cholera persisted until 2017 and resulted in 9000 deaths and upwards of 790,000 cases being reported [31].

The country of Dominican Republic which shares the island with Haiti had limited transmission of cholera due to the population having better access to safe water and improved sanitation and having five times higher gross domestic product than Haiti [27]. In the Dominican Republic, a total of 32,000 suspected cases and 500 deaths were reported due to the cholera outbreak [32]. This difference in the experiences of Haiti and Dominican Republic related to the morbidity, mortality, and persistence of outbreak shows the importance of safe water and improved sanitation and its association with poverty in widespread transmission and sustenance of the cholera outbreak after the introduction of a causative agent.

7.2 Peru Outbreak (1991–1995)

In late January 1991, the breakage of the water main was followed by the outbreak of cholera that exposed the rotting critical infrastructure and lack of safe water and improved sanitation in the population [25]. The cholera epidemic consists of two events of the introduction of agent and propagation due to environmental health conditions. Studies done in Peru in 1984–1985 had exposed the presence of poor quality water in most of the wells, springs, and surface water supplies [33]. The cholera outbreak in Peru and Latin American countries had an overwhelmingly high number of cases, morbidity, and mortality in the poor areas of the respective countries due to lack of basic critical infrastructure of safe water, sanitation, and sewer disposal [33]. The insufficient chlorination of the drinking water further contributed to the outbreak. In Peru, there were a total of 635,830 cases and 4418 deaths reported due to *V. cholerae* infection [34]. In regions of the Americas, a total of 1,075,372 cases and 10,098 deaths were reported due to cholera during this period [34]. A large part of these numbers was reported during the first two years of the outbreak.

Thus, both the recent outbreaks of cholera in the region of the Americas were associated with the lack of safe water and improved sanitation facilities that were exposed due to either natural disasters or human activity. The introduction of the agent in both the outbreaks was controversial but the propagation and sustenance of the outbreak were linked to the lack of basic critical infrastructure of public services. Further, the disparities related to the access of basic public services in these countries were related to economic status. The persistence of outbreak and spread to neighboring countries showed the importance of broader measures in preventing future outbreaks. The seasonal increase in the number of cases in these outbreaks was in line with seasonal variation seen in cases in endemic areas of the world with peaks at the end of monsoon seasons [35]. Higher flooding after monsoon increases the potential contact with contaminated water in low lying areas near rivers.

8 Methodology and Analysis

Crisis management begins well before any outbreak. It entails the application of scenario planning using predictive analytics. We applied the insights garnered from the analysis to a predictive model for Jamaica. In so doing, applied sensemaking is illustrated where cholera social vulnerabilities are revealed through the systems lens and leveraged in a predictive model. The case studies revealed systemic vulnerabilities in communities that contributed to the outbreaks. Based on the transmission pattern and previous outbreaks in cholera non-endemic regions, the risk factors for cholera outbreaks include poverty, safe water, improved sanitation, and proximity to rivers.

The country of Jamaica is divided into 14 administrative regions or parishes. For the preparation of potential cholera outbreak mitigation, a predictive risk assessment model was developed based on poverty indicators and closeness to water bodies in parishes. The data for poverty indicators were taken from Jamaica's Mapping of Poverty data and the number of water bodies in parishes was taken from google earth (Table 1). A predictive model with equal weightage to poverty and water bodies was developed as previous outbreaks had shown equal importance to both. Due to lack of data related to the availability of safe water and improved sanitation in the population of parishes, we only used poverty indicators for the model with an assumption that economically disadvantaged regions would be at higher risk of lack of improved water and sanitation. The poverty incidence below 0.125 was considered low, 0.125–0.250 was considered moderate while >0.250 was considered high. Similarly, having 1–3 water bodies was considered low, 4–6 was considered moderate and more than 6

Table 1 Factors and risk categories of cholera outbreak in Parishes of Jamaica

Parish	Poverty incidence data	Water bodies	Risk category
Hanover	0.104	2	Low
St. Elizabeth	0.244	1	Low
St. James	0.107	2	Low
Trelawny	0.179	3	Moderate
Westmoreland	0.167	4	Moderate
Clarendon	0.238	7	High
Manchester	0.243	1	Low
St. Ann	0.158	1	Low
St. Catherine	0.277	3	High
St. Mary	0.148	7	Moderate
Kingston	0.302	0	Moderate
Portland	0.153	9	High
St. Andrew	0.183	3	Moderate
St. Thomas	0.325	7	High

was considered high. The model helped in dividing the parishes into categories of low-moderate-high risk zones based on cholera outbreak risks (Table 1). This was followed by mapping of the parishes of Jamaica based on risk categories using ESRI ArcGIS software version 17.2.

9 Discussion

Jamaica is no stranger to natural disasters. The island of Jamaica is geographically located in an area that makes the country vulnerable to hurricanes. Fortunately, the island has not experienced a direct hit from a hurricane since Hurricane Gilbert in 1988 which will forever be etched in the memories of Jamaicans who experienced the Category 5 cyclone. Nevertheless, the 'near misses' that have been experienced since then have been significant to cause extensive damage to infrastructure and incur millions of dollars in economic losses. Next, Jamaica has not experienced a catastrophic earthquake since the September 12, 1907 earthquake which resulted in 1000 deaths, 9000 left homeless [36]. Prior to that was the June 07, 1692 earthquake that resulted in the sinking of most of the city of Port Royal, 5000 deaths, 3000 left homeless and a subsequent outbreak of yellow fever [36].

Jamaica has not had a cholera outbreak since 1851. Limited information is available on the first cholera outbreak in Jamaica; however, it is speculated that approximately 40,000 Jamaicans died at the time from the disease (10% of the population) [37].

During the period of 2020, Jamaica experienced (in addition to the COVID-19 pandemic) flooding due to heavy rainfall to several meteorological systems in the Caribbean which resulted in millions of agricultural losses, and significant damages to roads, houses and other communities. Jamaica also experienced moderate earthquakes that have gone unnoticed by several citizens—certainly nothing significant enough to cause major damages to infrastructure. There was only one exception so far, that is the 7.7 magnitude earthquake that occurred off the coast of Jamaica on January 28, 2020 approximately 80 miles off the north western coast of the island. News reports say that this earthquake was felt in neighbouring countries such as Cuba and Cayman and as far as Miami.

The bottom line is that Jamaica has not had its overall resilience as a country (to catastrophic events) audited by a major natural disaster in recent years. This is a blessing; however, it should lead to the questioning of whether the country is prepared for a catastrophic event as seen in its history. To many, it may seem as if the country is prepared, resilient and ready for natural disasters. However, these sentiments may exist out of ignorance. The island of Jamaica has not had a major catastrophic event in recent years for us to even imagine the extent to which a disaster of the sort would affect the country. Recall that sensemaking is defined as '…to how we structure the unknown so as to be able to act in it. Sensemaking involves coming up with a plausible understanding—a map—of a shifting world; testing this map with others through data collection, action, and conversation; and then refining, or abandoning, the map depending on how credible it is' [3]. Our mental models, (beliefs and assumptions about how the world works) shape what we see and how we label, categorize and make meaning thereby shaping our decisions and actions. Within the context of disaster and crisis management, sensemaking is about interpretation and action.

The world has seen examples of what a devastating disaster may look like in a worst-case scenario for Jamaica. Disasters in the region such as: the Haiti Earthquake (7 Magnitude) in 2010 followed by the cholera outbreak that lasted for 7 years, Hurricane Maria (Category 4) that hit Puerto Rico in 2017 and Hurricane Dorian (Category 5) that hit the Abaco Islands of the Bahamas in 2019 provide a glimpse of the catastrophic effects that a major natural disaster may have in the country. While all of these examples had different circumstances that contributed to the extensive damage and death that the world observed in horror (and cannot necessarily be used as a standard to critique Jamaica); it is important for us to learn lessons as a neighbouring Caribbean country—so that we can do the best that we can as a nation to prevent and mitigate (as best as possible) against the devastating effects that we have not seen in modern day Jamaica. This is especially so, seeing that the COVID-19 pandemic has already placed a significant toll on the health care sector and by extension the capacity of the country to respond to emergencies.

10 Predictive Model Results

The predictive model of anticipated hotspots for a cholera outbreak in Jamaica was based on the widely known fact that cholera is a disease that is predominantly associated with poverty. As such the model serves to highlight parishes across the Jamaica where cholera outbreaks may occur under circumstances discussed earlier in the chapter (the destruction of—and delay in repairing critical infrastructure as a result of a disaster). The risk map was created with 4 main assumptions (1) High rates of poverty in an area/community is affiliated with little or no resilience to catastrophic events. (2) Impoverished areas will have weak critical infrastructure compared to wealthier communities. (3) Overall resilience to a natural disaster such as a hurricane or earthquake will be less in parishes with a high poverty rate (4) The presence of surface water bodies (which in this case is mainly rivers), encourages the utilization of water from these sources which may be contaminated in the prolonged period after a catastrophic event. More information on the risk stratification was discussed in the methodology section.

It was observed that the following parishes—Clarendon, St. Catherine, Portland, and St. Thomas were considered high risk. These parishes are the parishes with the highest incidence of poverty in Jamaica (as seen in the 2019 report carried out by the Planning Institute of Jamaica, the Statistical Institute of Jamaica and the World Bank) and had the most water bodies where people may resort to as an alternative source of water. The parishes of Hanover, St. James, St. Elizabeth, Manchester and St. Ann were considered to be low risk due lower incidence rates of poverty and a

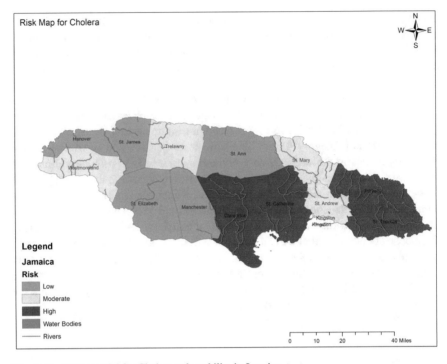

Fig. 1 Predictive model for Cholera vulnerability in Jamaica

small number of rivers being present. Westmoreland, Trelawny, St. Mary, Kingston and St. Andrew were considered to be medium risk (please refer to the methodology section for the procedure) (Fig. 1).

11 Limitations

Several limitations were encountered during this project. However, they are mainly associated with the lack of adequate water and sanitation data for Jamaica. We wanted to carry out the GIS mapping on a community level within these parishes. However, it was particularly difficult to get access to this data from public media such as publishable reports and peer-reviewed articles, government websites and governmental organizations—leaving the question whether or not the data exists at all. It is no secret that Jamaica does not have a uniformly distributed sewage system across the island outside of the metropolitan areas. However, even within these metropolitan areas, some individuals may have sanitation systems outside of a flushable toilet that is connected to a sewer main. Also, some communities may not be connected to a potable water line; or individuals within a particular community may rely on other water sources such as water catchment systems, the purchasing of potable water from trucks and wells for water. Therefore, the status of water and sanitation in communities across the island was unknown. The analysis would have been strengthened if this information was available. Other pieces of data that would have been helpful in this analysis include understanding local behaviour toward surface water sources i.e. understanding if individuals purify their water before drinking, cooking etc.

It is also very difficult to assess community resilience on a large-scale basis in a developing country like Jamaica. This due to the tightly coupled nature of some of the communities in the island. The country is a closely interwoven system, where a disruption in one 'connection' may have immediate effects in another community nearby. While resilience may exist in certain communities/neighbourhoods across the island, it is suspected (based upon participant observation) that Jamaica experiences *'community resilience'*—where a co-dependent relationship is formed between two or more neighbouring communities in order to boost the overall resilience of the systems combined. It is important to consider, because the resilience of Jamaican communities may not lie in and of themselves—it may be dependent on the other communities nearby. However, further research will be needed at a community level across the island to examine local systems and networks.

12 Sensemaking and Systems Solutions

The systemic risks and vulnerabilities inherent in the dynamics of cholera outbreaks have been illustrated in the case studies and contextual discussion. The cholera mitigation strategies that should be implemented needs to target the specific situation of each country. Sensemaking in this regard thereby requires the application of systems thinking to better understand the interdependencies and interconnectivity of societal systems. Suffice it to say that implementing these 'simple' prevention or mitigation strategies in the countries mentioned above will prove challenging under these conditions, due to the several dynamic variables in each country. Some of these variables that are integrated into the public health system include the economy, adequate disease surveillance, vaccination programs, increased access to safe water and sanitation infrastructure and a public health awareness campaign about the importance of clean water and WASH practices. Culture and education are also key factors pertaining to implementing mitigation strategies for cholera.

13 Surveillance and Vaccination

Surveillance in a cholera epidemic is very important because it supplies the necessary stakeholders with appropriate information needed to combat the outbreak. Cholera morbidity and mortality is "grossly underreported" globally. A revised 2017 position paper indicates that:

Approximately 1.3 billion people are at risk of cholera in endemic countries. An estimated 2.86 million cholera cases (uncertainty range 1.3 m – 4.0 m) occur annually in endemic countries. Among these cases, there are an estimated 95 000 deaths (uncertainty range: 21 000 – 143 000). [7]

While these are global statistics, they should be considered in this analysis as underreporting is synonymous with, "inadequate access to health care services among very poor and marginalized populations" [7]. The case study countries mentioned above have millions of vulnerable people fitting this description. Therefore, improved surveillance is one of the first steps that need to be taken in order to improve mitigation efforts in these countries. It will also help to improve efficiency and effectiveness of existing efforts to reduce the burden of the disease and prevention of future incidence rates.

In addition to improved disease surveillance, vaccination strategies would also make improvements. Literature provided by the WHO does indicate that research and development of the various cholera vaccines helps to reduce rates of incidence. However, it is also mentioned that these various cholera vaccines should be used to complement other prevention and mitigation strategies [7]. Surveillance and vaccinations as mitigation strategies go hand in hand in that adequate surveillance pinpoints where the vaccinations should be given a priority. The WHO fortunately has a stockpile of the oral cholera vaccine (OCV) which helps to provide inoculation support in emergencies [7]. Nevertheless, this is not the 'magic bullet' for reducing the effects of cholera. Water and sanitation infrastructure in addition to appropriate WASH practices remain the pillars for preventing waterborne diseases.

14   The Economy and Good Governance

The economy is the driving force for any form of major development within a country. A healthy economy will not only help to provide jobs but will also prevent certain dynamic pressures and unsafe conditions from occurring (as described in the pressure and release model [38]). In order for a country to be resilient to any form of hazards, their economy must be safeguarded with growth objectives. Good governance and a zero-tolerance approach to corruption also play a role in sustaining the economy through local industries and foreign investments. A mindset of inclusivity must also be entertained to ensure that basic human rights are available to all. These are some of the issues that preceded Zimbabwe's cholera outbreak in 2009. The political party at that time ZANU-PF affiliated with Mugabe simultaneously violated the basic human rights of several hundred thousand of Zimbabweans. Further neglect of their roles and responsibilities lead to a major economic and infrastructural collapse; contributing to the Zimbabwean dollar undergoing staggering 2,000,000% hyperinflation in 2008 [24]. Good governance is also important to implement legislation and policies to mitigate against the spread of disease. For example, Free-town in Sierra Leone is known for street vended food; however, data presented by Nguyen et al. [39] suggests that street vended food and drink may house harmful pathogens including *V. Cholera* [39]. Through good governance and thoughtful legislation with subsequent training and education; the average person can be aware of the dangers of dirty (although apparently clean) water.

Civil war and protest also have devastating long term effects for the sustainable development of a country's economy. While this was discussed briefly earlier on the chapter, it should be noted that Sierra Leone and Uganda are still recovering from the years of civil war that these countries faced. Therefore, while cholera is a pressing issue for global public health and while it is widely accepted that the pillars of cholera migration and prevention are safe water and sanitation infrastructure for all people. It appears from these case studies that successful cholera mitigation needs an indirect approach to enhance its effectiveness and sustainability. An approach that stems from economic growth and development that may have a spillover effect in other parts of these countries' portfolio.

It should also be noted that this indirect economic solution cannot be supported solely by individual countries. It would be unreasonable to expect that these poor countries can implement all the ideal infrastructure for themselves. It is going to take a collaborative effort from global partners (both governmental and private) to help the mitigation effort in these countries and others alike.

15 Technology

Technology also has a role to play in solving these complex public health situations. Perhaps, the conventional way of viewing the implementation of cholera mitigation strategies of water and sanitation infrastructure needs to be revisited. For example, rather than looking at the water and sanitation infrastructure from a macro-level; one may consider viewing the system as a collection of smaller, intimate and independent systems that work together for a common goal? While systems thinking and analysis is important in viewing how all the working parts are connected; scaling in and out on the system can be used to help build resilience one community at a time.

The technology aspect of the mitigation approach requires a level of creativity and innovation. As discussed earlier, all these countries are different. They all have similar needs, but the approach must to be tailored to specific ecological and environmental factors of each country. Technology along with education can also be used to incorporate the citizens of these high-risk communities in creating their own redemption—through education, empowerment and job creation while meeting their basic human rights and needs.

16　Climate Change and Extreme Weather Events

The vulnerable populations that are currently suffering due to cholera will be vulnerable under the impact of climate change. Climate change contributes to the propagation of cholera by increasing ideal temperatures for *V. Cholera* bacteria growth in addition to increasing the natural reservoirs due to rising sea levels for said bacteria. The seasonality of cholera outbreaks may also be affected by a change in weather patterns. Changing weather patterns may increase the amount of rainfall that a country experiences which may lead to flooding. Flooding contributes to cholera outbreaks by contaminating clean water supply with brackish water. While climate change is not directly associated with cholera outbreaks; it does appear to heighten several factors that contribute to an outbreak and should be taken into consideration.

Natural disasters are also not directly associated with an immediate cholera outbreak but play a huge role in the deterioration of critical infrastructure. The length of time it takes to rebuild after the disaster does contribute to the risk of a cholera outbreak [19]. This speaks towards the importance of safe-guarding critical infrastructure and building resilient communities to help mitigate against future outbreaks. It also relates to what was discussed earlier in regard to the economy. This is also a lesson for regions such as the Americas and in particular considering the focus of this chapter, also a lesson for Jamaica.

17　Recommendations and Conclusions

This chapter should stimulate the reader to consider the pre-existing conditions, hazards and vulnerabilities that exists within our society/ critical infrastructure and then, utilize a system thinking approach to troubleshoot areas of improvement (whether individually or on a larger scale) where targeted interventions can be implemented, to improve our resilience as individuals and by extension a country located in a disaster-prone region.

This analysis highlighted Jamaican parishes that should be on the radar for cholera outbreaks after a catastrophic event based on the information provided earlier in this chapter. It also revealed the need for further research on this topic. Access to reliable data is key. It is important from a public health perspective to thoroughly understand the water and sanitation conditions of a country, within both urban and rural areas as well as the supporting societal and public health infrastructure. A part of this is the requirement for policy research supporting societal infrastructure and emergency management. This research endeavour would require a collaborative effort with relevant public health organizations and institutions, the Jamaican government and any other key stakeholder (including concerned citizens). The systems view highlights that a new perspective for societal infrastructure is required that integrates not only the elements of WASH but also poverty, governance and education.

Systems thinking and predictive analytics figure prominently to support sense-making in this solution space. As seen with the COVID-19 pandemic and equally applicable to cholera, 'Evidence-based communications and policy outreach as part of an overall advocacy strategy is critical to create a shift in stakeholder and public opinion, expedite decisions and mobilize resources to highlight an issue or gap in programs. This can be a strategic and deliberate process that helps build awareness, visibility and public momentum behind an issue, integrating and coordinating the independent efforts of academicians, epidemiologists, public health officials, civil society, media and policymakers into actions for positive change' [40].

References

1. Deen J, Mengel M, Clemens JD (2020) Epidemiology of cholera. Vaccine 38(2020):A31–A40
2. Weick KE (1995) Sensemaking in organizations. Sage, Thousand Oaks, CA
3. Ancona DL (2011) SENSEMAKING *Framing and acting in the unknown*. In: *Handbook of teaching leadership*. Sage Publications, Inc., Thousand Oaks, CA, pp 3–20
4. Nowling D, Seeger MW (2020) Sensemaking and crisis revisited: the failure of sensemaking during the Flint water crisis. J Appl Commun Res 2020 48(2):270–289. https://doi.org/10.1080/00909882.2020.1734224
5. Weick KE (1993) The collapse of sensemaking in organizations: The Mann Gulch disaster. Admin Soc 38(4):628–652
6. De Magny GC, Colwell RC (2009) Cholera and climate: a demonstrated relationship. Retrieved from https://www.ncbi.nlm.nih.gov/pmc/articles/PMC2744514/
7. World Health Organization (2017, Aug 21) Cholera position paper. Retrieved from https://www.who.int/immunization/policy/position_papers/cholera/en/
8. World Health Organization (2019, June 14) Drinking-water. Retrieved from https://www.who.int/news-room/fact-sheets/detail/drinking-water
9. Ogliastri E, Zúñiga R (2016) An introduction to mindfulness and sensemaking by highly reliable organizations in Latin America. J Bus Res 69(2016):4429–4434
10. Jackson MC (2003) Systems thinking: creative holism for managers. Wiley, West Sussex
11. Masys AJ, Izurieta R, Reina M (ed) (2019) Global health security: recognizing vulnerabilities, creating opportunities. Springer Publishing
12. Edson R (2008) Systems thinking. Applied: a primer. ASysT Institute. https://www.asysti.org/OurModules/SharedDocs/HistoryDocView.aspx?ID=373
13. Masys AJ (ed) (2016) Applications of systems thinking and soft operations research in managing complexity. Springer Publishing
14. Senge P (1990) The fifth discipline: the art and practice of the learning organization. Doubleday Currency, New York
15. Ackoff R (1994) Systems thinking and thinking systems. System Dyn Rev 10(2–3):175–188
16. Peters DH (2014) The application of systems thinking in health: why use systems thinking? Health Res Policy Sys 12:51
17. Massaro E, Ganin A, Perra N, Linkov I, Vespignani A (2018) Resilience management during large-scale epidemic outbreaks Nat Sci Rep 8:1859
18. Knap AH, Rusyn I (2016) (2016) Environmental exposures due to natural disasters. Rev Environ Health 31(1):89–92
19. Julta A, Khan R, Colwell R (2017, Jan 27) Natural disasters and cholera outbreaks: current. Retrieved from https://www.researchgate.net/profile/Rakibul_Khan4/publication/312952087_Natural_Disasters_and_Cholera_Outbreaks_Current_Understanding_and_Future_Outlook/links/5a438a410f7e9ba868a58684/Natural-Disasters-and-Cholera-Outbreaks-Current-Understanding-and-Future-Outlook.pdf

20. Cambaza E, Mongo E, Anapakala E, Nhambire R, Singo J, Machava E (2019) Outbreak of Cholera Due to Cyclone Kenneth in Northern Mozambique, 2019. Int J Environ Res Publ Health. 16:2925. https://doi.org/10.3390/ijerph16162925
21. Bwire G, Malimbo M, Maskery B, Kim YE, Mogasale V, Levin A (2013) The burden of cholera in Uganda. PLoS Neglected Trop Dis 7(12). https://doi.org/10.1371/journal.pntd.0002545
22. The Combat Genocide Association (n.d.) Uganda 1971–1985. Retrieved from https://combatgenocide.org/?page_id=91
23. The World Bank (2010) Water supply and sanitation in Sierra Leone: turning finance into services for 2015 and beyond. Retrieved from https://www.wsp.org/library/water-supply-and-sanitation-sierra-leone-turning-finance-services-2015-and-beyond
24. Cuneo CN, Sollom R, Beyrer C (2017, Dec) The cholera epidemic in Zimbabwe, 2008–2009: a review and critique of the evidence. Retrieved from https://www.ncbi.nlm.nih.gov/pmc/articles/PMC5739374/
25. Cerda R, Lee PT (2013) Modern cholera in the Americas: an opportunistic societal infection. Am J Public Health 103(11):1934–1937. https://doi.org/10.2105/AJPH.2013.301567
26. Guthmann JP (1995) Epidemic cholera in Latin America: spread and routes of transmission. J Trop Med Hyg 98(6):419–427
27. Tappero JW, Tauxe RV (2011) lessons learned during public health response to cholera epidemic in Haiti and the Dominican Republic. Emerg Infect Dis 17(11):2087–2093. https://doi.org/10.3201/eid1711.110827
28. Centers for Disease Control and Prevention (2010) Update: cholera outbreak—Haiti, 2010. MMWR Morb Mortal Wkly Rep 59(45):1473–1479
29. Centers for Disease Control and Prevention (2010) Update: outbreak of cholera—Haiti, 2010. MMWR Morb Mortal Wkly Rep 59(48):1586–1590
30. Tauxe R, Seminario L, Tapia R, Libel M (1994) The Latin American epidemic. In: Vibrio cholerae and Cholera. Wiley, pp 321–344. https://doi.org/10.1128/9781555818364.ch21
31. Ministre De La Sante Publique Et De La Population (2016) Rappot De Cas [Case Report]. https://mspp.gouv.ht/site/downloads/Rapport%20Web%2020%2008%202016%20Avec%20Courbes%20departementales.pdf
32. Government of Canada (2017) Cholera in Dominican Republic and Haiti (Travel health notice). https://web.archive.org/web/20170612162441/, https://travel.gc.ca/travelling/health-safety/travel-health-notices/111
33. Reiff FM (1992) Cholera in Peru (World Health, p 2). World Health Organization. https://apps.who.int/iris/bitstream/handle/10665/52728/WH-1992-Jul-Aug-p18-19-eng.pdf?sequence=1&isAllowed=y
34. Pan American Health Organization (1995) Cholera in the Americas. Epidemiol Bull 16(2). https://www.paho.org/english/sha/epibul_95-98/be952choleraam.htm
35. Hashizume M, Faruque ASG, Wagatsuma Y, Hayashi T, Armstrong B (2010) Cholera in Bangladesh: climatic components of seasonal variation. Epidemiology (Cambridge, Mass.), 21(5):706–710. https://doi.org/10.1097/EDE.0b013e3181e5b053
36. Jamaica Information Service (2017, Jan 10) Major earthquakes in Jamaica. https://jis.gov.jm/information/get-the-facts/major-earthquakes-jamaica/
37. Thompson E (2010, Nov 3) Jamaica's 1851 Cholera outbreak. The Gleaner.
38. Hammer CC, Brainard J, Innes A et al (2019) (Re-)conceptualising vulnerability as a part of risk in global health emergency response: updating the pressure and release model for global health emergencies. Emerg Themes Epidemiol 16:2
39. Nguyen VD, Sreenivasan N, Lam E, Ayers T, Kargbo D, Dafae F et al (2014) Cholera epidemic associated with consumption of unsafe drinking water and street-vended water—Eastern Freetown, Sierra Leone, 2012. Am J Trop Med Hyg 90(3):518–523. https://doi.org/10.4269/ajtmh.13-0567
40. Nayyar A, Privor-Dumm L (2020) Cholera control and prevention: role of evidence-based advocacy and communications. Vaccine 38(2020):A178–A180
41. Planning Institute of Jamaica, Statistical Institute of Jamaica, & World Bank (2019) Mapping poverty indicators—consumption based poverty in Jamaica

42. Piarroux R, Barrais R, Faucher B, Haus R, Piarroux M, Gaudart J, Magloire R, Raoult D (2011) Understanding the cholera epidemic, Haiti. Emerg Infect Dis 17(7):1161–1168. https://doi.org/10.3201/eid1707.110059
43. https://news.un.org/en/node/1039381/mozambique-cyclones-a-wake-up-call-to-boost-resistance-un-weather-agency-2

Sensemaking and Security: How Climate Change Shapes National Security

Ross Prizzia

Abstract Climate change has evolved from an environmental issue, to an energy problem, to a security threat with an impending sense of urgency. It is increasingly recognized as having national security implications with multifaceted national security risks. Climate change increases the intensity and frequency of extreme weather events that threaten nations' territorial integrity and sovereignty. For example, rising sea levels cause both internal displacement within nations and climate-change refugees across national borders. Some of the climate change threats facing the world in 2020 include droughts, wildfires, floods and global pandemics. Each one has the potential to increase competition and conflict around the world, creating more instability than ever. With many military bases susceptible to these threats, the changing climate has been a focus of military planners for many years. The National Defense Strategy (NDS), the primary planning document that guides decision making within the United States military, emphasizes climate change as a major factor in an increasingly complex security environment. Climate change will likely increase instability and insecurity within already vulnerable regions, particularly in regard to its destructive impacts, as rising seas infiltrate coastal bases, hurricanes batter installations, and wildfires infringe on training ranges and impact readiness. This chapter provides an analysis of climate change as a national security issue in the context of climate science and climate change's multifaceted security effects with an emphasis on the most disaster-prone regions of Asia and the Pacific.

Keywords Climate change · Sensemaking · National security

1 Introduction

Climate change has evolved from an environmental issue, to an energy problem, to a security threat with an impending sense of urgency. It is increasingly recognized as having national security implications with multifaceted national security risks.

R. Prizzia (✉)
Public Administration and Disaster Preparedness, University of Hawaii-West Oahu, Kapolei, USA
e-mail: rprizzia@hawaii.edu

© The Author(s), under exclusive license to Springer Nature Switzerland AG 2021
A. J. Masys (ed.), *Sensemaking for Security*, Advanced Sciences and Technologies for Security Applications, https://doi.org/10.1007/978-3-030-71998-2_6

Climate change increases the intensity and frequency of extreme weather events that threaten nations' territorial integrity and sovereignty. For example, rising sea levels cause both internal displacement within nations and climate-change refugees across national borders. Some of the climate change threats facing the world in 2020 include droughts, wildfires, floods and global pandemics. Each one has the potential to increase competition and conflict around the world, creating more instability than ever. With many military bases susceptible to these threats, the changing climate has been a focus of military planners for many years.

The National Defense Strategy (NDS), the primary planning document that guides decision making within the United States military, emphasizes climate change as a major factor in an increasingly complex security environment. Climate change will likely increase instability and insecurity within already vulnerable regions, particularly in regard to its destructive impacts, as rising seas infiltrate coastal bases, hurricanes batter installations, and wildfires infringe on training ranges and impact readiness.

The transnational character of climate-related security risks often extends beyond the capacity of national governments to respond adequately. This chapter provides an analysis of climate change as a national security issue in the context of climate science and climate change's multifaceted security effects with an emphasis on the most disaster-prone regions of Asia and the Pacific.

The release of a series of think tank reports on climate change and national security highlights how the U.S. national security apparatus has moved to incorporate climate change into its strategic planning. Climate change's direct threats to the homeland and indirect threats to the country's overseas interests are factored into how to effectively align means and ends, from military preparation to humanitarian aid, at a time of constrained resources.

One report by a panel of security professionals—"A Security Threat Assessment of Global Climate Change: How Likely Warming Scenarios Indicate a Catastrophic Security Future"—analyzed the security implications of two future warming scenarios (near term: 1–2 °C and medium-long term: 2–4+ °C). The report identifies major threats, including heightened social and political instability, risks to U.S. military missions and infrastructure, as well as other security institutions at both warming scenarios and across all regions of the world. The primary findings and recommendations include [20]:

Findings

1. A near-term scenario of climate change in which the world warms 1–2 °C/1.8–3.6 °F over pre-industrial levels by mid-century would pose 'High' to 'Very High' security threats. A medium-to-long-term scenario in which the world warms as high as 2–4+ °C/3.6–7.2 °F would pose a 'Very High' to 'Catastrophic' threat to global and national security. The world has already warmed to slightly below 1 °C compared to pre-industrial temperatures.
2. At all levels of warming (1–4+ °C/1.8–7.2+ °F), climate change will pose significant and evolving threats to global security environments, infrastructure, and institutions.

3. While at lower warming thresholds the most fragile parts of the world are the most at risk, all regions of the world will face serious repercussions. High warming scenarios could bring about catastrophic security impacts across the globe.
4. These threats could come about rapidly, destabilizing the regions and relationships on which U.S. and international security depend.
5. Climate change will present significant threats to U.S. military missions across all of its geographic areas of responsibility (AORs), as well as to regional security institutions and infrastructure that are critical for maintaining global security.

Recommendations

1. Mitigating climate change risks in order to avoid severe and catastrophic security futures requires quickly reducing and phasing out global greenhouse gas emissions. This calls for the world to achieve net-zero global emissions as soon as possible in a manner that is swift, ambitious, safe, equitable, and well-governed.
2. The world must "climate-proof" environments, infrastructure, institutions, and systems on which human security depends and also rapidly build resilience to current and expected impacts of climate change, future-oriented investments in adaptation, disaster response, and peacebuilding.
3. In the United States there is an immediate need to prioritize, communicate, and respond to climate security threats, and to integrate these considerations across all security planning.

In August of 2020, the U.S. House and Senate passed the Fiscal Year 2021 National Defense Authorization Act (NDAA), which includes the following declarations:

1. Climate change poses a direct threat to the national security of the United States.
2. A requirement that the Department of Defense (DoD) prioritize its vulnerabilities and send to Congress a list of its most vulnerable installations.
3. An expansion of existing authorities to incorporate climate change considerations.
4. Improvements to building codes and a requirement for DoD to conduct resilience planning at each of its installations.

The recommendations help to identify the unique vulnerabilities at each military installation. The 2020 bills in the U.S. House and Senate continued support of the DoD's work on climate resilience. Some key provisions include [2]:

- Requirement to update the 2014 Climate Change Adaptation Roadmap. Congress specifically indicated that this update shall include a strategy and implementation plan of the DoD to address the current and foreseeable impacts of climate change on DoD missions, geopolitical/strategic environments, infrastructure inside and outside the U.S., and civilian dependencies such as supply chains and strategic transportation nodes. Once completed, this document is likely to guide the next several years of DoD climate activities.

- Establishment of a National Academies Climate Security Roundtable to create a mechanism for climate science stakeholders to provide information to the Climate Security Advisory Council (CSAC). The CSAC was established within the Intelligence Community in the 2019 bill and was inspired by the Climate Security Crisis Watch Center [20].
- Improvement of water management and security on installations.
- Direction to the U.S. Coast Guard to assess its vulnerabilities and among other requirements to identify the ten sites most vulnerable to climate change impacts. This assessment addresses the requirement passed for DoD in the Fiscal Year 2018 NDAA.
- Authorization for the DoD to fund projects that improve military installation resilience even when they are outside the borders of the installation or on land the DoD does not control.
- Requirement for an assessment of the impact of permafrost thaw on DoD assets and operations.
- An assessment building upon DoD's 2018 vulnerability report that would focus exclusively on extreme weather vulnerability of military installations and combatant commander requirements.
- Allocation of additional planning and design funds ($50 million) to support military installation resilience projects.

These proposed provisions reflect a continuation of the progress made by the U.S. legislative branches and DoD on climate change and security considerations evident in both Republican and Democratic administrations for the past two decades.

The rapid rate of climate change and its implications for global climate security present unprecedented risks. Technological developments have provided climate models and predictive tools that enhance our ability to anticipate and mitigate some of these complex risks and the creation of 'Responsibility to Prepare and Prevent (R2P2)' will provide a framework for managing climate security risks. This framework encompasses and addresses what is known about climate security risks, what gaps exist in governing these risks, and how to close this global governance gap. The main climate security governance gaps identified follow [25]:

Gap 1: The Right Information. Currently, there is no standardized global hub for climate security information to inform coherent international policy actions that can address climate security risks or offer realistic future projections in the field.

Gap 2: The Right People. Addressing climate security risks is hampered by a gap between climate change messengers and the security audiences necessary to take actions to address climate security risks. There is also a lack of institutionalized leadership devoted to addressing climate security risks within the global community.

Gap 3: The Right Time. Inadequate global governance mechanisms for aligning international climate policy actions with the interests of national policy actions add to climate security risks.

The proposal for the establishment of an international R2P2 Climate Security Governance Framework is comprised of three institutional principles [25]:

Principle 1: Assessment and Anticipation. Standardized, aggregated and credible global climate security assessments including climate security futures aimed at aiding coherent international action. This should start with the establishment of an International Climate Security Assessment Panel (ICSAP) to oversee a regular climate security assessment report.

Principle 2: Elevation and Translation. Leadership by senior, globally-respected security practitioners who translate climate security information for global security decision-makers and issue regular recommendations for international action. This should start with the appointment of a UN Permanent Representative for Climate and Security to lead a new Climate Security Center (CSC) to fulfill this goal.

Principle 3: Coordination and Alignment. International climate security coordination mechanisms for aligning international climate change policy with international security policy as they relate to climate security risks. This should begin with the establishment of an intergovernmental "Climate Security Coordination Mechanism (CSCM)" at the UN level to lead this coordination [25].

2   The World Climate and Security Report 2020

In February 2020, the Expert Group of the International Military Council on Climate and Security (IMCCS) released its inaugural "World Climate and Security Report 2020" at the Munich Security Conference (MSC), the annual and influential gathering of senior international security and military leaders [9]. The IMCCS is a group of senior military leaders, security experts, and security institutions across the globe dedicated to anticipating, analyzing, and addressing the security risks of a changing climate. The group was founded and is administered by the Center for Climate and Security (CCS), an institute of the Council on Strategic Risks (CSR), in partnership with the French Institute for International and Strategic Affairs (IRIS), the Hague Centre for Strategic Studies (HCSS), and the Planetary Security Initiative of the Netherlands Institute of International Relations. These organizations form the Expert Group of the IMCCS [9].

The IMCCS report is written from the vantage point of international military and security experts providing a global overview of the security risks of a changing climate. It identifies opportunities for addressing these risks and recommends "climate-proofing" international security, including infrastructure, institutions and policies, as well as reducing major emissions to avoid significant-to-catastrophic security threats [9].

While there has been progress over the past decades with military and security institutions increasingly analyzing and incorporating climate change risks into their assessments, plans, and policies, the IMCCS report urges that more must be done as risks are more extreme and urgent. This contributed to the report's "Key Risks and Opportunities" findings which describe five significant or higher risks to global security under current circumstances and three key opportunities as a path forward for global security cooperation on climate change [9].

Key Risks:

1. Water insecurity a global security risk: Climate change-exacerbated water insecurity is already a significant driver of instability, and according to 93% of climate security and military experts surveyed for this report, it will pose a significant or higher risk to global security by 2030.
2. All regions face increase in climate security risks, not just the fragile/poor: Although fragile regions of the world suffer the most severe and catastrophic security consequences of climate change, all regions face significant or higher security risks due to the global nature of the risks. Of the climate security and military experts surveyed for this report, 86% perceive climate change effects on conflict within nations to present a significant or higher risk to global security in the next two decades.
3. Military institutions are increasingly concerned about climate risks: As reinforced by the 31 nations represented in the International Military Council on Climate and Security (IMCCS), a growing number of national, regional and international security and military institutions are concerned about and planning for climate change risks to military infrastructure, force readiness, military operations, and the broader security environment.
4. Climate mitigation, adaptation and resilience efforts are critically needed to avert the significant security consequences of climate change: However, some proposed solutions such as geoengineering could present negative second-order effects to global security if not implemented carefully.
5. Rising authoritarianism, sharpened global competition and national agendas are hampering the needed cooperation among nations to address the security risks of climate change.

Key Opportunities:

1. National, regional, and international security institutions and militaries around the world should advance robust climate resilience strategies, plans and investments, especially regarding climate implications for water and food security and their associated effects on stability, conflict and displacement.
2. Military and security institutions should demonstrate leadership in managing climate security risks and resilience and encourage governments to advance comprehensive emissions reductions and adaptation investments to avoid security disruptions. Military organizations can also lead by example by taking advantage of the significant opportunities to adopt lower carbon energy sources, and reducing other greenhouse gases beyond carbon dioxide.
3. Climate-proofing development assistance for vulnerable nations which are likely hotspots of instability and conflict, as well as climate-proofing policies affecting those regions, should be a priority for conflict prevention. Assistance

should be aimed at climate resilience challenges such as water security, food security, and disaster preparedness.

UN researchers claim that the first twenty years of this century witnessed a "staggering" rise in climate disasters resulting in 7348 recorded disaster events worldwide causing the death of approximately 1.23 million people and $2.97 trillion in losses to the global economy [7]. As a global superpower with military forces deployed around the world, the interests of the U.S. and its allies will be impacted by a changing climate, especially in certain 'hot-zones" [17]. Among the most impacted of these hot-zones is the Asia Pacific Region.

3 Climate Change Security in South Asia

South Asia, including India, Bangladesh, Pakistan, Afghanistan, Iran, Maldives, Sri Lanka, Nepal and Bhutan is one of the world's poorest and most illiterate regions with the highest levels of underdevelopment [8]. It is home to nearly one fourth of the world's population, and it is among the world's most militarized regions, with India and Pakistan sharing a long border and holding nuclear capabilities [13].

Due to the geography of the region, South Asia is hard hit by the effects of climate change. Poverty and underdevelopment are key factors in South Asia's vulnerability to climate change because they reduce people's capacity to cope with large-scale disasters as well as to adapt for future disruptions [18].

The coastal zones of South Asia contain about 40% of the economic activity of the region and most of its critical economic infrastructure, including ports, fishing infrastructure and mangrove areas. Approximately 80% of the land mass of Bangladesh consists of a delta made up of a complex system of flood plains [18]. An increase of severe weather, like cyclones, puts these areas at tremendous risk, ultimately affecting fragile government structures as the economy suffers.

Economic losses from natural hazards differ widely among countries, even when accounting for the intensity of the disaster. For South Asia with its high levels of poverty, illiteracy and underdevelopment the economic impact of climate change can be devastating. A study funded by the World Bank Group and Global Facility for Disaster Reduction and Recovery (GFDRR) found that disasters' impact on gross domestic product (GDP) is 20 times higher in developing countries than in industrialized nations [28]. Moreover, for every person in wealthy countries who died in a disaster in the last 50 years, almost 30 individuals died in poor countries [22].

South Asia's large land mass with diverse populations and ecosystems challenge emergency, risk and security management. There are some generalizations about areas in South Asia that could be helpful in planning. These generalizations include:

- With the increase of natural disasters and endemic poverty in South Asia, the region will be severely affected by climate change.
- The U.S. may need to provide disaster assistance to the millions who may be displaced by increased extreme weather.

- India is strategically important to the United States as a rising power. The tense relationship between India and Bangladesh and the potential for a large climate-induced migration must be carefully monitored.
- The U.S. may be required to provide mediation or troop support if tensions between India and Bangladesh turn to violence.
- The U.S. has strategic military interests in Pakistan and Afghanistan with numerous military installations, including Diego Garcia air base that could be compromised by climate change.

Taking into account these generalizations optimizes emergency and security planning by military and security planners.

4 Climate Change Security in the Asia Pacific Region

Security in the Asia–Pacific Region is of great importance to the United States as conveyed in the 2015 U.S. government report by the Center for Climate and Security, "The U.S. Asia–Pacific Rebalance, National Security and Climate Change". The report overview states that the United States is in the early stages of what it characterizes as an "Asia–Pacific rebalance", and that the United States intends to reorient its foreign policy and national security posture in the Asia–Pacific region with its large populations, growing economies, and "strategic choke-points" like the South China Sea and a number of rising powers [26].

The report describes the Asia–Pacific region as "one of the most vulnerable to the effects of climate change, with a growing coastal population, rising seas, numerous critical waterways fed by glaciers, threatened island states, increased drying, and projections of severe water insecurity," and in the near future, "the effects of climate change are likely to both shape, and be shaped by, the U.S. role in the Asia–Pacific" [26].

The report concludes that "if the U.S. is to engage constructively in the region—building and broadening alliances, helping advance regional security and prosperity in the face of potentially catastrophic change, and advancing U.S. national security interests—it will have to seriously consider how climate change affects the region, how the U.S. can help advance the climate resilience of the region's diverse nations, and how the U.S. will adapt strategically to a changed security environment" [26].

The South Pacific alone has roughly 25,000 islands with many islands like Singapore and islands in the Philippines and Indonesia that have extremely high population densities. Since the 1950s, natural disasters, including cyclones, earthquakes and droughts have occurred more frequently than ever before in the South Pacific; 10 of the 15 most extreme natural disasters have occurred in the first 15 years of the twenty-first century. Population growth and economic instability mean that these disasters have had larger human impacts.

Nearly 40% of all the disasters triggered by natural hazards in the world occur in Asia, and 88% of the people affected reside in this region. Of the total number of people affected in Asia, the People's Republic of China (PRC) and India account for just over 40%, reflecting their population size and land mass. However, after accounting for population size and land area, Bangladesh, Philippines, India, the PRC, Maldives, and Japan, in this order, have been the top six countries affected since 2000.

Both developing and developed countries of Asia and the Pacific are highly exposed to natural disasters. Asia–Pacific nations experience more natural disasters than any other region in the world. Between 2014 and 2017, nations in this region were affected by 55 earthquakes, 217 storms and cyclones, and 236 cases of severe flooding impacting 650 million people and causing the deaths of 33,000 people [27]. Of the 12 disasters with the highest death tolls across the world since 1980, nine occurred in Asia. In 2011, 80% of global disaster-related economic losses occurred in the Asia Pacific region. The losses caused by these disasters were immense not only in terms of human lives but also in terms of property destroyed. A conservative estimate of the average annual direct economic damage due to disasters in countries of Asia and the Pacific in the period 2001–2011 was US$ 60 billion [22].

One example of the severity and destructive impact of an extreme weather event is Super Typhoon Haiyan which struck in November 2013. Super Typhoon Haiyan made landfall in the Philippines, wreaking havoc and massive loss of life and property. It is considered the most powerful typhoon of all time. Even a year after it struck millions of people suffered its devastating effects, with little or no access to basic needs such as food, clean water, and healthcare. Oxfam warned that governments needed greater investment in climate and disaster-resilient development and more effective assistance for those at risk to prevent loss of lives and homes to extreme weather.

With the threat that Super Typhoon Haiyan-scale disasters could fast become the norm, not the exception, Asian states have begun to adopt policies and programs to reduce the risks and to increase adaptive responses to climate change impacts such as extreme weather and rising sea levels. The staggering costs of the impacts of climate change are economically burdensome to the victim countries. With 4.3 billion people or 60% of the global population, Asia has borne almost half the estimated economic cost of all disasters over the past 20 years, amounting to around US$ 53 billion annually [14]. The report by Oxfam urged the Association of Southeast Asian Nations (ASEAN) to create a regional resource base to help member states carry out projects to adapt to climate change impacts and manage risk [14].

According to the National Intelligence Council (NIC), by 2025, "unprecedented economic growth, coupled with 1.5 billion more people, will put pressure on resources—particularly energy, food, and water—raising the specter of scarcities emerging as demand outstrips supply" [5]. Climate Change natural disasters only exacerbates already tenuous relationships among nations competing for scarce resources. Global climate change strategically impacts vital U.S. security interests because it affects defense, diplomacy and economics, cause an increase in frequency of disaster relief responses by the U.S. military, and threaten the regional stability of U.S. allies [4].

The U.S. has strategic and economic interests in South-East Asia, specifically in terms of China's rise as a global economic power. The U.S. military has numerous military installations in South East Asia and the Pacific. Scattered with low-lying island nations, the region is highly vulnerable to rising sea levels. A severe extreme weather event, or a series of them, could put the military installations and personnel at risk as well as have consequences on the economic strength of the region, thereby affecting the global economy.

Global climate variability and change increase the frequency and severity of natural disasters and security risks in the Asia–Pacific region, threatening the fabric of life for people in the region. Vital health, environmental and social dimensions including access to clean water, food production, and the sustainability of ecological systems and the urban built environment are at risk. Severe weather is predicted to become more frequent and destructive in the Asia–Pacific region and warming trends are expected to bring new security challenges.

By 2070, 9 out of 10 of the world's most populous cities will be in Asia, including low-lying mega-regions like Bangkok, Manila and Ho Chi Minh City. These cities are growing rapidly and their lack of infrastructure and high density make them particularly vulnerable to flooding and storms. Higher tides and storm surges wash away underground soil that supports buildings and infrastructure and seasonal precipitation in the region is expected to rise between 2 and 6% by 2050, threatening flood prone areas in Asian megacities. Floods are by far the most frequently occurring disasters in Asia and claim the highest numbers of victims.

Because of the built-in momentum in the climate system resulting from past emissions and the limited capacity of the oceans to absorb and neutralize the harmful impacts, the physical risks posed by climate change will continue to grow into the next century even if a dramatic reduction in greenhouse gas emissions is achieved. Rather than slowing, climate change will likely accelerate as emissions continue to grow. Recent modeling results suggest a possible warming of 5.2 °C by 2100 [24].

Even conservative estimates predict that the rising temperatures and changing ocean levels in the Asia–Pacific Region will lead to significant socio-economic, environmental and security risks. Sea rise for coastal cities may be particularly damaging, especially as people and population densities continue to increase in flood plains and coastal areas of the Asia–Pacific region. Higher temperatures, rising seas and a more energetic hydrologic cycle are expected to contribute to more intense storms, droughts, crop failures and food insecurity. This has serious implications for national security in the Asia Pacific Region including the mass migration of "climate refugees" [19] across international borders and increased conflict among nations competing for scarce resources, particularly among upstream and downstream nations in Asia.

Therefore, emergency managers, security professionals and governments must initiate climate adaptation and mitigation measures to help protect communities in the Asia–Pacific region. This is crucial in mitigating the impacts of a changing climate which act as accelerants of instability by multiplying problems like water scarcity, food shortages, and overpopulation. These problems increase the region's vulnerability by fueling natural resource competition and underdevelopment while burdening civilian and military institutions around the world, including the U.S. military [23].

5 Climate Change Security in the Pacific Islands

Natural disasters, like cyclones, can engulf entire cities and small islands, creating large-scale chaos and damage. Repeated events like these cause the contamination of fresh water supplies forcing a redistribution of fisheries. Each flood and extreme storm increase the salinity of fresh water and alters the biodiversity on land and sea, with direct effects on the economy, food and water security [24]. As the climate changes, flooding is more likely as storms become increasingly unpredictable. Flooding from extreme events increases the likelihood of spreading diseases like cholera, typhus, and dengue fever. The spread of disease is likely to be exacerbated by the warm weather environment and the potential for standing water, common to Pacific islands.

Less fresh water, more coastal erosion, and degraded coral reefs are among the impacts climate change is already having on Hawaii and other Pacific islands associated with the United States, according to the "Third National Climate Assessment" report [21]. More than 300 scientists contributed to the study confirming that extreme weather events linked to climate change, including heat waves, heavy downpours, floods, and droughts, have become more frequent and intense throughout the United States and Pacific islands. These events are disrupting people's lives and hurting the economy. Highlights from the section on Hawaii and the Pacific include [21]:

- Decreasing rainfall in low-lying areas, combined with a rise in sea levels that push seawater into aquifers, will limit the availability of fresh water.
- Rising sea levels combined with increased storm runoff will increase coastal flooding and erosion, damaging coastal ecosystems, infrastructure, and agriculture.
- A warming ocean will increase coral bleaching and disease outbreaks on coral reefs.
- Rising temperatures and reduced rainfall in some areas will put native plants and animals at greater risk for extinction.
- Pacific Islanders will find it increasingly difficult to sustain their traditional ways of life as climate change forces them to leave coastal areas.

Using mid-range scenarios, in which a rise in the sea level of 40 cm is envisaged by the 2080s, the number of people threatened worldwide from coastal flooding is projected to more than double to 200 million [15]. Results show that sea-level rise is an ongoing and accelerating process with a high likelihood of becoming a grave danger to coastal communities on Pacific islands. In the Pacific Ocean, meltwater is expected to constitute a long-term threat of sea-level rise in the second half of the twenty-first century, with thermal expansion of the upper ocean posing the greatest immediate challenge.

The citizens of some Pacific island states and deltaic coasts do not have the luxury of retreating inland from the coast and may face involuntary relocation. For example, in the Pacific nation of Tuvalu, a ring of nine Polynesian islands, several thousand people have already left for other nations because of rising seas. Displaced refugees from low-lying areas could provide the human reservoir for the spread of disease, including malaria [3]. Basic health, economic, social and other human needs are left unmet. Accordingly, businesses, non-governmental organizations and the public sector have an obligation to evaluate the impacts of sea level rise on Pacific Islanders and to propose innovative solutions to mitigate these effects.

Climate-related factors threaten to worsen existing fragile situations beyond the tipping point for many Pacific Island governments, even those that appear stable. In response, developed nations have begun to consider the best ways to assist low lying island states as the impacts of sea-level rise and climate change begin to take their toll on families, communities and nations of the Pacific islands. The international aid agency, Oxfam, released a blueprint [14] for Australia's new engagement with Pacific nations. Recommendations include: reduce greenhouse gas emissions by at least 95% by 2050; develop renewable energy alternatives; provide financial support funding to help Pacific nations adapt to rising sea levels; and assist communities displaced by the results of climate change, including governance arrangements and preparations for forced immigration.

6 Military Planning for Humanitarian Assistance and Disaster Relief (HADR)

Military planning for Humanitarian Assistance and Disaster Relief (HADR) capacities requires military planners to consider the four essential activities of disaster and emergency management—Reduction (mitigation), Readiness (preparedness), Response, and Recovery. This is especially important with the increasing likelihood of multiple hazards and converging risks in a climate changed future. One example is the back-to-back disasters in July and August 2018 that affected over five million people, when floods resulting from monsoon rains, tropical storms and a dam collapse on a Mekong tributary affected parts of Myanmar, Laos, Cambodia, Vietnam and the Philippines, while an earthquake and multiple aftershocks struck Indonesia's Lombok Island in West Nusa Tenggara [1].

When disaster strikes, governments can seek the support of the military in the Response and Recovery phases. The benefits of using the military start with reconnaissance operations in a disaster area to identify specific capabilities required; minimize suffering, loss of life, property damage, and especially in the case of climate-related disaster, degradation to the environment; organize a command center; and deploy disaster relief into the relevant disaster area. With its unique capabilities, including trained and disciplined personnel, the military can deploy rapidly to support first responders and other stakeholders tasked with tackling complex disaster operations. Effective disaster response is an important part of strengthening resilience to climate impacts and climate security risks [1].

The role of the military in natural disaster relief operations is increasing in the Indo-Asia Pacific region. Across South and Southeast Asia, the military traditionally plays an important role in disaster response, and will continue to do so as climate change drives more frequent and severe extreme weather events.

Ensuring HADR capabilities that can meet the scale of extreme disasters is critical in South and Southeast Asia. The region's high exposure and vulnerabilty to extreme weather events and disastrous economic losses are accentuated by the concentration of economic infrastructure along coasts and in coastal megacities, jeopardizing economic growth that underpins stability for many countries in the region. Adequate disaster response systems are a key component of the social contract between governments and citizens. Inadequate provision of basic needs and services can prompt discontent and changes in government or strengthen the legitimacy of non-state actors.

The military is integral to disaster response systems across South and Southeast Asia. With their surge capacity and command-and-control structure, the militaries in the region are the first de facto national responders in the emergency phase of a disaster event in coordination with civilian authorities [1]. It is likely that the role of the military in natural disaster relief operations will expand as climate change drives more frequent and extreme natural disasters.

7 Adapting Bases and Troops at Risk

Military leaders must contend with climate change impacts on bases, forces and equipment. Hurricanes Florence and Michael in 2018 and heavy inland flooding in the spring of 2019 caused an estimated US$ 10 billion in damage to Marine Corps Base Camp Lejeune in North Carolina, Tyndall Air Force Base in Florida and Offutt Air Force Base in Nebraska. Scientists widely agree that climate change is making storms like these larger, more intense and longer-lasting.

Threats to other bases particularly those located along U.S. coastlines, such as the giant naval station at Norfolk, Virginia are bound to grow as sea-levels rise and major storms occur more frequently. Rising temperatures generate other challenges. In Alaska, many military facilities are at risk of collapse or damage as the permafrost on which they sit begins to thaw. In California, wildfires burn on or near key bases. Extreme heat also poses a health risk to soldiers, who must often carry heavy loads during sunlit hours, and danger to the safe operation of helicopters and other mechanical equipment [10].

Recognizing these dangers, the armed forces are acting to reduce its vulnerability. They have built seawalls at Langley Air Force Base adjacent to Norfolk Naval Station and are relocating sensitive electronic equipment at coastal bases from ground level to upper stories or higher elevations. The Defense Department also is investing in renewable energy, including solar power and biofuels. By the end of 2020, the armed forces expect to generate 18% of on-base electricity from renewables, up from 9.6% in 2010. They plan to increase that share substantially in the years ahead.

As this outlook suggests, human communities face far greater risks from climate change in the short term than scientists' habitat loss projections into 2100 and beyond may suggest. Vulnerable societies are crumbling under the pressure of extreme climate effects and the scale of chaos and conflict is certain to grow as global temperatures rise [10]. This is already happening in some communities, such as Norfolk, Virginia where base commanders and local officials have found common ground in addressing the area's extreme vulnerability to sea-level rise and hurricane-induced flooding. The armed forces continue to plan for conventional conflicts abroad, while recognizing that climate change will affect their ability to perform their combat duties.

8   The Need for Climate Change Security Strategy

The COVID-19 pandemic has rapidly reshaped global economic, security and diplomatic conditions, halted the global economy and intensified tensions between Washington and Beijing. 2020 was expected to be an extraordinary year for climate, as the United States and other governments planned to announce new, bold goals and establish action plans to address climate change and related security issues.

The anticipated climate plans were derailed in large part by the Covid 19 pandemic. This is particularly the case for the United States, which has experienced the highest number of infections and deaths, a staggering number of lost jobs and trillions in increased national debt. The United States is at a critical crossroad and must move quickly to respond to the growing geopolitical security threats stemming from climate change and intensified by global threats like Covid 19 [6].

Recognizing the effects of a changing climate that is currently and in the future will continue to be a national security issue impacting Department of Army installations, operational plans, and overall missions, the Army released a memo directing installations to plan for energy and climate resilience efforts by identifying the installations' vulnerability to climate-related risks and threats. This memo is consistent with Department of Defense April 2020 guidance of Master plans for major military installations. To address climate risks and threats, the Army Climate Resilience Handbook (ACRH) guides Army planners through a process to systematically assess climate exposure impact risk and incorporate the resulting knowledge and data into existing installation planning processes such as straegic master plans.

The 2019 NDAA Section 2805 defines climate resilience as the "anticipation, preparation for, and adaptation to utility disruptions and changing environmental conditions" [16]. Using this understanding of climate resilience, the ACRH guides Army planners through a four-step risk-informed planning process to develop a Climate Vulnerability Assessment that identifies:

(1)   the installation's climate resilience goals and objectives
(2)   how exposed the installation is to current nuisance and extreme weather events and to projected future climate impacts
(3)   how sensitive infrastructure, assets, mission, and readiness are to these impacts and how difficult adapting to these threats may be
(4)   a list of potential measures that can be used to improve an installation's preparedness and resilience.

A key element of the ACRH process is the Army Climate Assessment Tool, (ACAT), that provides climate change impact information at the installation, command, and headquarters levels that is specifically developed for use in the screening-level assessment of climate exposure risk. ACAT includes reports that identify those installations that have the greatest exposure to identified climate change impacts. This handbook is divided into two main sections, an ACRH overview and an in-depth explanation of the four-step ACRH process. The report utilizes a simulated Army base as an example to give a cohesive and comprehensive understanding of the outputs from each step. Appendices provide resilience measures, additional climate change and ACAT information, a short user's guide for the ACAT, and a glossary of terms [16].

Throughout the world, DoD installations are exposed to the risks of climate change, jeopardizing U. S. security. By anticipating and reducing exposure to future climate change threats, the Army can lessen climate impacts to missions and operations and protect its real property investments. The ACRH is intended to help Army planners in this effort [16].

In the final analysis, and from an international perspective, it should be noted that most government agencies rely on reports provided by the Intergovernmental Panel on Climate Change (IPCC) scientific baseline data to inform analysis of the effects of climate change on national security. The pace and scope of climate change as a threat to national security are imminent and undeniable [12]. To ignore the relationship between climate change and national security puts nations, regions, and the world at risk for insurmountable human and ecological danger and loss.

References

1. Climate and security in the Indo-Asia Pacific part of the world climate and security report 2020 briefer series (2020, July). Retrieved from https://imccs.org/wp-content/uploads/2020/07/Climate-Security-IndoAsia-Pacific_2020_7.pdf. Accessed 11 May 20
2. Conger J (2020) Climate security in the 2021 U.S. National Defense Authorization Act. Center Clim Secur. Retrieved from https://climateandsecurity.org/2020/08/climate-security-in-the-2021-u-s-national-defense-authorization-act/. Accessed 10 Feb 2020
3. Constable A (2016) Climate change and migration in the Pacific: options for Tuvalu and the Marshall Islands. Reg Environ Change. Retrieved from https://www.researchgate.net/publication/305111515_Climate_change_and_migration_in_the_Pacific_options_for_Tuvalu_and_the_Marshall_Islands. Accessed 15 Sept 20. https://doi.org/10.1007/s10113-016-1004-5

4. Defense Science Board (2011) Report of the defense science board task force: trends and implications of climate change for national and international security. Retrieved from https://www.fas.org/irp/agency/dod/dsb/climate.pdf. Accessed 10 Feb 20
5. Fingar CT (ed) (2009) Global trends 2025: a transformed world. DIANE Publishing
6. Goodman S, Bergenas J (2020) The United States needs a natural security strategy to regain our global leadership. Morning Consult. Retrieved from https://morningconsult.com/opinions/the-united-states-needs-a-natural-security-strategy-to-regain-our-global-leadership/. Accessed 10 Feb 20
7. Homeland Security (2020) Climate challenges: "staggering" rise in climate emergencies in last 20 years. Homeland Security News Wire. Retrieved from https://www.homelandsecurityne wswire.com/dr20201012-staggering-rise-in-climate-emergencies-in-last-20-years. Accessed 20 Sept 2020
8. Human Development Centre (Islamabad, Pakistan), Mahbub ul Haq Human Development Centre (2007) Human development in South Asia. Oxford University Press
9. IMCCS (2020) The world climate and security report 2020 February 2020. A product of the expert group of the international military council on climate and security. Retrieved from https://climateandsecurity.org/wp-content/uploads/2020/02/world-climate-sec urity-report-2020_2_13.pdf. Accessed 25 Sept 20
10. Klare M (2020) A military perspective on climate change could bridge the gap between believers and doubters. The Conversation. Retrieved from https://theconversation.com/a-mil itary-perspective-on-climate-change-could-bridge-the-gap-between-believers-and-doubters-128609. Accessed 10 Feb 2020
11. MIT News (2009) The climate response—Government leaders take action, evaluate vulnerabilities due to climate change
12. Melton M (2018) Climate change and national security, part I: what is the threat, when's it coming, and how bad will it be? Retrieved from climate change and national security, Part I: what is the threat, when's it coming, and how bad will it be? Lawfare (lawfareblog.com). Accessed 27 Nov 20
13. Najam A (2003) The human dimensions of environmental insecurity: some insights from South Asia. Environ Change Secur Project Rep 9:59–74
14. National Intelligence Council (2012, Feb 2) Global water security. Intelligence Community Assessment: iii. Oxfam. (2014, Nov). 10 natural disasters that shook the world in 2014. Oxfam Report
15. Patz J, Kovats S (2002) Hotspots in climate change and human health. BMJ 325:1094–1098
16. Pinson AO, White K, Moore S et al (2020) United States Army Corps of Engineers (USACE). Retrieved from https://www.preventionweb.net/publications/view/73906. Accessed 10 Feb 20
17. Podesta J, Ogden P (2008) The security implications of climate change. Was Q 31(1):115–138
18. Rahman AA, Chowdhury ZH, Ahmed AU (2003) Environment and security in Bangladesh. In: Najam A (ed) Environment, development, and human security: perspectives from South Asia. University Press of America, Lanham, pp 103–128
19. Salem S, Rosencranz A (2020, July) Climate refugees in the Pacific. Retrieved from https://elr.info/news-analysis/50/10540/climate-refugees-pacific. Accessed 11 Jan 20
20. The Center for Climate and Security (2020, Feb 18) A security threat assessment of global climate change how likely warming scenarios indicate a catastrophic security future. An International Security Threat Assessment of Two Warming Scenarios conducted by U.S. National Security, Military and Intelligence Professionals. Retrieved from https://climateandsecurity.org/wp-content/uploads/2020/03/a-security-threat-assessment-of-climate-change.pdf. Accessed 10 Feb 2020
21. Thompson D (2014) Hawaii already seeing effects of climate change. Honolulu Mag
22. UNISDR-AP (2011) United Nations Office for Disaster Risk Reduction—Regional Office for Asia and Pacific (UNISDR-AP). In: Climate change and disaster management at the crossroads: climate change adaptation and disaster risk reduction in Asia and the Pacific, 162p
23. US Department of Defense (2010, Feb) Quadrennial defense review report. Retrieved from https://dod.defense.gov/Portals/1/features/defenseReviews/QDR/QDR_as_of_29JAN10_1 600.pdf. Accessed 10 Feb 2020

24. Weir T, Virani Z (2011) Three linked risks for development in the Pacific Islands: climate change, natural disasters and conflict. Clim Dev 3(3):193–208
25. Werrell C, Fermia F (2019) The responsibility to prepare and prevent: a climate security governance framework for the 21st century. Retrieved from https://climateandsecurity.org/the-responsibility-to-prepare-and-prevent-a-climate-security-governance-framework-for-the-21st-century/. Accessed 10 Feb 20
26. Werrell C, Fermia F (eds) (2015) The U.S. Asia-Pacific rebalance, national security and climate change. Center Clim Secur (1025 Connecticut Ave., NW, Suite 1000, Washington, DC)
27. Wood J (2018) Why Asia-Pacific is especially prone to natural disasters. World Econ Forum. Retrieved from https://www.weforum.org/agenda/2018/12/why-asia-pacific-is-especially-prone-to-natural-disasters/. Accessed 11 Jan 20
28. World Bank (2014) World bank disaster risk management overview. Retrieved from https://www.worldbank.org/en/topic/disasterriskmanagement/overview. Accessed 10 Feb 20

Importance of the Humanitarian—Development—Peace Nexus to Make Sense for Security Some Thoughts and Examples from Palestine

Hildegard Lingnau

Abstract The world is far from achieving security. The cold war came to an end 30 years ago and impressive progress has been made in terms of development (MDGs, SDGs), but regarding security it seems that the overall patterns have changed without tangibly improving security. So the question is: How to make better sense for security? From a humanitarian and development perspective security is not to be achieved by military means. Security starts with food security and ends with human security/peace. In between, many other dimensions, especially the development dimension come into play. This chapter discusses the importance of the Humanitarian-Development-Peace nexus to make sense of security.

Keywords Humanitarian-Development-Peace nexus · Food security · Human security

1 Progress on Many Fronts, but not Regarding Peace and Security

The world is far from achieving security. The cold war came to an end 30 years ago and impressive progress has been made in terms of development (MDGs, SDGs), but regarding security it seems that the overall patterns have changed without tangibly improving security (Sect. 1).

So the question is: How to make better sense for security? From a humanitarian and development perspective security is not to be achieved by military means. Security starts with food security and ends with human security/peace. In between, many other dimensions, especially the development dimension come into play. Section 2 introduces and discusses relevant concepts.

Acting WFP Country Director in Palestine and Adjunct Professor at Siegen University, currently teaching at Bethlehem University. The views expressed herein are those of the author(s) and do not necessarily reflect the views of the World Food Programme.

H. Lingnau (✉)
Siegen University, Siegen, Germany
e-mail: hildegard.lingnau@web.de

© The Author(s), under exclusive license to Springer Nature Switzerland AG 2021
A. J. Masys (ed.), *Sensemaking for Security*, Advanced Sciences and Technologies for Security Applications, https://doi.org/10.1007/978-3-030-71998-2_7

Section 3 zooms into the Humanitarian—Development—Peace (HDP) nexus—a concept which emerged in 2016—and spells out the triple nexus concretely by looking into humanitarian work/food assistance (3.1.), development work (3.2) and human security/peace (3.3) and respective contextual particularities. The section shows how important a joined-up approach would be in practice, but also illustrates how difficult it is to translate concepts into reality/practice.

Section 4 explains the specificities and limitations of the country case.

The contribution concludes (Sect. 5) by stating that despite little progress over the five years since the concept was agreed the nexus is still relevant and worth pursuing: The talk needs to be walked now for countries to effectively benefit from the famous triple HDP nexus concept.

2   How to Tackle This Challenge? How to Make Better Sense for Security? Some Conceptual Thoughts

The **Sustainable Development Goals** (SDGs) agreed in September 2015 and succeeding the Millennium Development Goals (MDGs) as the world's agenda for development for the next 15 years include for the first time a goal on security: SDG 16 ("Peace, justice and strong institutions") with the following targets:

- Significantly reduce all forms of violence and related death rates everywhere.
- End abuse, exploitation, trafficking and all forms of violence against and torture of children.
- Promote the rule of law at the national and international levels and ensure equal access to justice for all.
- By 2030, significantly reduce illicit financial and arms flows, strengthen the recovery and return of stolen assets and combat all forms of organised crime.
- Substantially reduce corruption and bribery in all their forms.
- Develop effective, accountable and transparent institutions at all levels.
- Ensure responsive, inclusive, participatory and representative decision-making at all levels.
- Broaden and strengthen the participation of developing countries in the institutions of global governance.
- By 2030, provide legal identity for all, including birth registration.
- Ensure public access to information and protect fundamental freedoms, in accordance with national legislation and international agreements.
- Strengthen relevant national institutions, including through international cooperation, for building capacity at all levels, in particular in developing countries, to prevent violence and combat terrorism and crime.
- Promote and enforce non-discriminatory laws and policies for sustainable development.

According to the UN progress report towards the SDGs [1] conflict, insecurity, weak institutions and limited access to justice remain a great threat to sustainable development. Millions of people have been deprived of their security, human rights and access to justice. In 2018, the number of people fleeing war, persecution and conflict exceeded 70 million, the highest level recorded by the Office of the United Nations High Commissioner for Refugees in nearly 70 years. The pandemic is potentially leading to an increase in social unrest and violence, which would greatly undermine the world's ability to meet the targets of SDG 16.

The **SIPRI Yearbook** 2020 [2] confirms this: "This 51st edition of the SIPRI Yearbook provides evidence of an ongoing deterioration in the conditions for international stability. This trend is reflected in the continued rise in military spending and the estimated value of global arms transfers, an unfolding crisis of arms control that has now become chronic, and increasingly toxic global geopolitics and regional rivalries. There also remains a persistently high number of armed conflicts worldwide, with few signs of negotiated settlements on the horizon."

And so does the **OECD States of Fragility Report** [3] by stating

- That more countries are experiencing violent conflict than at any time in nearly 30 years,
- That 77% of extremely poor people live in fragile contexts and
- That 90% of humanitarian assistance is going to protracted crises.

Peace and security is more than the absence of war and violence.

But insecurity is not only increasing, patterns are also changing as **Dieter Senghaas** already predicted in 1988 [4]: "Der zweite Entwicklungstrend (…) besteht in der Herausbildung einer Art von "Chaos-Macht". (…) Aus tendenziell chaotischen Situationen (…) entsteht Verhinderungsmacht" (S. 170f).[1] The World Bank and UN study "Pathways for peace" (2018) states that "between 2010 and 2016 alone, the number civilian deaths in violent conflicts doubled" and that "many more civilian deaths result from indirect effects of conflict, such as unmet medical needs, food insecurity, inadequate shelter, or contamination of water" ([5]: xx).

This points to a different way of looking at security. From a humanitarian and development perspective security it is obvious that security is not about the absence of physical violence, ceasefires, negotiated peace deals only. Security starts with food security and ends with human security. In between many other dimensions and many other means come into play beside military dimensions and means. Security is also about political participation and democratic regimes, food security and many other "ingredients" (see Sect. 3.3).[2]

The Nobel Peace Prize Laureate for Economic Sciences 1998, **Amartya Sen**, demonstrated empirically in"Democracy as Freedom" [7] that"no famine has ever taken place in the history of the world in a functioning democracy". Sen explained that democratic governments "have to win elections and face public criticism, and

---

[1]"The second development trend (…) consists in the emergence of a kind of "chaos power". (…) Situations that tend to be chaotic will give rise to powers of destructive opposition (…)".

[2]Also see IASC for concepts of negative and positive peace ([6]: 6).

have strong incentive to undertake measures to avert famines and other catastrophes. There has not been a large-scale loss of life since 1947."

The **Nobel Peace Prize awarded to the World Food Program** in 2020 also clearly acknowledges that food (security) is key for security: "The link between hunger and armed conflict is a vicious circle: war and conflict can cause food insecurity and hunger, just as hunger and food insecurity can cause latent conflicts to flare up and trigger the use of violence. We will never achieve the goal of zero hunger unless we also put an end to war and armed conflict. The Norwegian Nobel Committee wishes to emphasise that providing assistance to increase food security not only prevents hunger but can also help to improve prospects for stability and peace" [8].

Many steps …

Over the last decades a whole range of concepts have been developed how to best frame and tackle the challenge with humanitarian and development means. Among the first one was the "**New Deal for Engagement in Fragile States**" (developed by the International Dialogue on Peacebuilding and Statebuilding[3] and signed by more than 40 countries and organizations at the 4th High Level Forum on Aid Effectiveness 2011 in Busan, Korea), an agreement between fragile and conflict affected states, international development partners and civil society to improve development policy and practice in fragile states. Countries committed themselves to pursuing more political ways of working to address the root causes of conflict and fragility and to channelling investments in fragile states in line with development effectiveness principles.

In 2015, the **2030 Agenda for Sustainable Development** explicitly referred to the relationship between development and peace: "There can be no sustainable development without peace and no peace without sustainable development" ([10]: 4). It was also acknowledged that humanitarian emergencies threaten "to reverse much of the development progress made in recent decades" ([10]: 6). The SDGs can therefore be seen as one of the first and most important efforts to encompass these three dimensions ([11]: 4).

At the 5th global meeting of the International Dialogue [12], the "**Stockholm Declaration on Addressing Fragility and Building Peace in a Changing World**" called for

- Addressing the root causes of fragility, conflict and violence, in a political way grounded in indigenous contexts,
- Using development aid in more innovative ways by increasing country programmable aid, strengthening public financial management systems and

---

[3]The International Dialogue is the first forum for political dialogue to bring together countries affected by conflict and fragility, development partners, and civil society. The International Dialogue is composed of members of the International Network on Conflict and Fragility (INCAF), the g7 + group of fragile and conflict-affected states, and the Civil Society Platform for Peacebuilding and Statebuilding (CSPPS) [9].

- Working more closely together with the different actors (humanitarian, development and UN peace building actors).

At the first **World Humanitarian Summit** 2016 in Istanbul a **Grand Bargain** was agreed which is the name for a set of 51 commitments to reform humanitarian financing (inter alia to deliver an extra billion dollars over five years, to provide 25% of global humanitarian funding to national and local responders, to provide more un-earmarked and multi-year funding to ensure greater predictability and continuity, to cut bureaucracy through harmonised reporting requirements etc.).

At this occasion the UNSG launched the **Agenda for Humanity** [13] in which the UN commits "to collective outcomes based on comparative advantages" and "to reduce fragmentation of international assistance into unmanageable numbers of projects and activities" ([13]: 1).

The related **New Way of Working** [14] aims to frame the work of humanitarian and development actors and places the notion of "collective outcomes" centre stage [14], which are defined as "the result that development and humanitarian actors (and other relevant actors) want to have achieved at the end of 3–5 years" ([14]: 6).

... leading to HEDP nexus concept ...

The **Humanitarian—Development—Peace (HDP) nexus** is the most recent concept which developed over the last years and gained momentum since the UN Secretary General António Guterres expanded the double nexus to the triple nexus and made it a central element of the UN agenda [13, 15, 16]. There is no agreed definition or document[4] but an overall understanding that the humanitarian, development and peacebuilding efforts are complementary and potentially mutually reinforcing. As a consequence, the respective "silos" need to be broken and divides to be transcended ([10]: 10) to have greater and mutually reinforcing impacts.

... and beyond

In 2018, a major breakthrough was achieved in establishing an international policy and legal framework for addressing conflict-related food insecurity: The UN Security Council passed **resolution 2417**, which highlights the two-way relationship between food insecurity and conflict: It confirms the prohibition of the use of hunger as a weapon of war (it was blocking access to food to force civilian populations to surrender or leave) and recognized that the world will never be able to eliminate hunger unless there is peace.

The World Bank—together with the UN—further developed the notion and the approach in the publication "**Pathways for peace**—Inclusive approaches to preventing violent conflict" [5]. The study starts by stating that the resurgence of violent conflict in recent years is an obstacle to achieving the Sustainable Development Goals by 2030 and therefore "seeks to improve the way in which domestic policy processes interact with security, diplomatic, justice and human rights efforts to

---

[4]According to Howe "there has been significant confusion over what the triple nexus means in both conceptual and practical terms and how this approach concretely contributes to progress on the Sustainable Development Goals" ([11]: 1).

prevent conflicts from becoming violent" (World Bank 2019: 2). To understand what works, it reviews the experience of different countries and institutions and comes to the following results, conclusions and recommendations ([5]: xviiif):

1. "Violent conflict has increased (…).
2. The human and economic cost of conflicts around the world requires all of those concerned to work more collaboratively. (…)
3. The best way to prevent societies from descending into crisis (…) is to ensure that they are resilient through investment in inclusive and sustainable development. For all countries, addressing inequalities and exclusion, making institutions more inclusive (…) are central to preventing the fraying of the social fabric that could erupt into crisis. (…)
4. The primary responsibility for preventive action rests with the state (…).
5. Exclusion from access to power, opportunity, services, and security creates a fertile ground for mobilizing group grievances to violence (…).
6. Growth and poverty alleviation are crucial but alone will not suffice to sustain peace. (…) It also means seeking inclusive solutions through dialogue, adapted macroeconomic policies, institutional reform in core state functions, and redistributive policies.
7. Inclusive decision making is fundamental to sustaining peace at all levels (…). Fostering the participation of young people as well as organisations, movements, and networks that represent them is crucial. Women's meaningful participation in all aspects of peace and security is critical (…)."

The study "presents the evidence to support a renewed focus on prevention" (p. 277) and also makes a strong business case for prevention showing that

- Prevention is economically beneficial (average net savings of 5 to 70 billion USD per year),
- Prevention is also good for the international community as it saves on post conflict humanitarian assistance and peacekeeping interventions (which are much more expensive than preventive action),
- The benefits of prevention increase over time, while the costs fall ([5]: 3).

In 2019, the OECD DAC members agreed on **recommendations on the nexus**.[5] Their main ambition is "shifting from delivering humanitarian assistance to end need. This will be critical in reducing the humanitarian caseload" ([17]:3). As a consequence, the priorities are seen as follows: "prevention always, development whenever possible, humanitarian action when necessary" ([17]: 3). The 11 principles which were agreed are the following:

1. Undertake joint risk-informed, gender-sensitive analysis on root causes and structural drivers of conflict (…)
2. Provide appropriate resourcing (…)
3. Utilise political engagement (…) to prevent crises, resolve conflicts and build peace (…)

---

[5]In June 2020, WFP adhered to the OECD DAC recommendations on the HDP nexus.

4. Prioritise prevention (…), investing in development whenever possible while ensuring immediate humanitarian needs continue to be met (…)
5. Put people at the centre, tackling exclusion and promoting gender equality (…)
6. Ensure that activities do no harm, are conflict sensitive to avoid unintended negative consequences and maximise positive effects (…)
7. Align joined-up programming with the risk environment (…)
8. Strengthen national and local capacities (…)
9. Invest in learning and evidence (…)
10. Develop evidence-based humanitarian, development and peace financing strategies (…)
11. Use predictable, flexible, multi-year financing (…).

Most recently, at the WFP Executive Board meeting on 16.11.2020 (HLP **"Breaking the silos"**), global leaders from the African Union, the European Union, the International Monetary Fund, the World Bank and the United Nations committed themselves again to "enhanced co-convening across the Humanitarian-Development-Peace nexus in fragile, conflict and violence affected countries" (Outcome 2). Panelists indicated their "willingness to examine how best to further enhance country-level planning and coordination, with particular attention on fragile, conflict and violence affected countries building on the recommendations from the *Pathways for Peace* report and the OECD DAC Recommendation on the Humanitarian-Development-Peace Nexus."

3   The Importance of the Humanitarian—Development—Peace (HDP) Nexus and Some Concrete Ideas and Examples How to Operationalize Activities at the Nexus

In 2018 WFP stated that the "lack of clarity about what the linked humanitarian, development, and peace action looks like in practice poses a challenge to moving forward jointly" ([18]: 7). To tackle this challenge and to illustrate that work at the Humanitarian—Development—Peace (HDP) nexus is not only desirable but also feasible in practice, the following section will move from theory to reality on the ground by looking into humanitarian work/food assistance (3.1.), development work (3.2) and human security/peace (3.3) and respective contextual particularities. The section shows how important a joined-up approach would be in practice, but also illustrates how difficult it is to translate concepts into reality/practice. The country context is Palestine. Palestine is a very specific country and case (mainly because it has never been granted independence and is suffering from occupation). The specificities and limitations of this country case will be discussed in the following section (Sect. 4).

A Nexus Initiative in Palestine

In 2019, some members of the UN Country Team (the Office of the Humanitarian/Resident Coordinator, OCHA, UNDP, WFP and FAO) embarked on developing a triple nexus approach for Palestine and conceptualizing UN work at the nexus [19]. The following challenges—problems and potential solutions -were identified: Data collection, needs assessments and analyses, targeting, planning and programming, fund raising and financing, implementation, monitoring and reporting are all done separately and there are too many coordination for a which do not help but hamper nexus work.

Data collection, needs assessments and analyses are done individually by all agencies. Beside or on top of this, a Humanitarian Needs Overview is done by the humanitarian actors, a Common Country Analysis (CCA) and a UN Development Assistance Framework (UNDAF) are done by the development actors. Not to forget the multitude of spontaneous data collection exercises and surveys done by all actors. A nexus approach should create a common database and establish a secretariat to update data regularly—ideally in close cooperation with the Palestinian Statistical Office. A nexus approach should also assess needs and analyse data jointly including the humanitarian, development and peace dimensions.

Targeting: UN agencies use data from different sources and target beneficiaries in different ways, very much driven by their mandates and headquarter policies. A nexus approach should work with the government, especially its Statistical Office, to identify and target people most in need. These data should be used by the government and by all agencies.

Planning and programming: take place for humanitarian work (Humanitarian Response Plan) and for development work (UN Development Assistance Framework, UN Sustainable Development Cooperation Framework). Both are not well funded so in practice all agencies do whatever they manage to get funding for. A nexus approach should bring both "silos" together and have joined-up planning and programming (an "Integrated Strategic Framework") which defines collective outcomes.[6] This would require the redefinition of some mandates (especially OCHA's mandate) since having a dual or triple mandate (such as WFP) is the exception, not the rule.

Fund raising and financing: Joint resource mobilisation efforts (such as the Humanitarian Response Plan which is done by OCHA) get some, but not enough attention and funding.[7] As a consequence, all agencies keep on fundraising individually and get most of their funding bilaterally. A nexus approach should develop and implement collective outcomes and joint financing and resource mobilisation strategies to attain these.

---

[6]The Office for the Coordination of Humanitarian Affairs (OCHA) suggests a 7-step-by-step-approach: 1. Determine the "lay of the land" for collective outcomes, 2. Ensure decisive leadership and strong support capacity, 3. Effective analysis of/optimize existing data to define priority areas, 4. Articulate collective outcomes, 5. Operationalizing collective outcomes—planning and programming, 6. Aligning resources and finances, 7. Monitoring and accountability (see [20]).

[7]Human Response Plans are heavily underfunded. The same is true for the UN Central Emergency Response Fund (CERF).

Implementation: Implementation plans and implementation are done by all agencies on their own, sometimes based on or linked to the Human Response Plan respectively the UN Development Assistance Framework. The UN Sustainable Development Cooperation Framework Guidance however recommends adopting a Cooperation Framework so that agencies do not have to prepare separate plans. This would save time, improve targeting and delivery and enhance knowledge sharing.

Monitoring: The same is true for monitoring. A nexus approach by contrast should collect common data for a common monitoring, review and evaluation plan.

Reporting is done separately, too. A nexus approach should replace HRP, UNDAF and individual agencies reports with a One UN country results report.

Coordination: While humanitarian work is officially being coordinated within humanitarian clusters (food security, education, health and nutrition, protection, WASH, shelters) and the Humanitarian Country Team led by Humanitarian Coordinator, an Inter-Cluster Coordination Group (ICCG) and OCHA, development work is coordinated in a whole range of fora: sector working groups (co-chaired by government authorities), the UN Country Team (Resident Coordinator, Heads of Agencies, Deputies, UNDAF working groups, UNDAF Steering Committee, SDG taskforce, other UN thematic groups on gender, communication etc.). These are relevant fora for exchange, but they do not help to craft, draft and implement work at the HDP nexus. Most agencies coordinate vertically with their headquarters and funders, but not horizontally with UN sister agencies and other humanitarian and development actors. A nexus approach should have a Nexus Consortium and Secretariat which would encourage and (if needed) enforce collective outcomes.

All in all, the UN colleagues who started thinking about work at the nexus agreed that the current situation implies a whole range of serious disadvantages (duplication of efforts, competition for resources, sub-optimal outcomes etc.) and that this is not only ineffective and inefficient but also against the true spirit of the UN. The relevant papers were tabled, discussed and approved at an annual retreat of the UN country team, but never implemented. Very obviously, it proved difficult for some actors to move out of their existing business as usual. But in principle, the ambition still persists: A nexus approach is a great opportunity to demonstrate that the UN system is nimble, relevant and effective. At a time when the multilateral system in general and the UN in particular is increasingly being called into question it would be good to do so.

Humanitarian, Development and Peace Principles

One of the more difficult challenges to tackle is that all 3 worlds ("silos") have their own principles. The humanitarian principles (humanity, neutrality, impartiality and independence[8]), the development cooperation principles (harmonization, alignment and results orientation[9]), peace principles (do no harm et al.). Humanitarian actors are for example concerned that work at the triple nexus will run the risk of politicisation

---

[8]See UN General Assembly resolution 46/182.

[9]For the full list of development cooperation principles see: https://www.oecd.org/dac/effectiveness/.

and instrumentalization of humanitarian assistance. They are worried that there could be tension between the need to align with government priorities to support a nationally led development and the need to not be perceived as supporting one party to the conflict over others (see ([18]: 6). This could result in a loss of neutrality in the eyes of local actors. But if the HDP is seen as a spectrum[10] it opens up options for action instead of limiting them. No matter what none of these principles should be used as a pretext to stay in the comfort zone of established silos.[11]

A Focus on Prevention

Last but not least work at the nexus should have an overall focus on prevention. If root causes (not only symptoms) are addressed and if joint and holistic scenarios are done, the risk of a deterioration of a humanitarian and/or development situation and the risk of an escalation of violence may be reduced. This is emphasized by Pantuliano [21]: "For the UN to become more effective at preventing crises, three key things need to happen. The UNSG needs to ensure that there are credible, high-quality and accountable leaders who will introduce and drive a culture of prevention both at HQ and in country operations. The Secretariat, Agencies, Funds and Programmes must overhaul systems and processes to pool existing financial and human resources more effectively towards prevention priorities. Finally, the UN must act as a catalyst for others on crisis prevention—emphasizing partnerships while reflecting on its limitations and comparative advantages on a case-by-case basis."

3.1 Humanitarian Work/Food Security

The World Food Programme is the world's largest humanitarian organisation providing food assistance to food insecure people in need. WFP is "committed to participating in humanitarian-development joint needs assessments, combined data analysis, and planning and programming processes to deliver better outcomes to people, moving beyond meeting needs to ending needs" [22]. Humanitarian food assistance is the core work of WFP reaching approximately 100 million people in 88 countries of the world.

After steadily declining for over a decade, global hunger is currently on the rise again. According to the State of Food Security and Nutrition in the World 2020 report (see [23]), hunger affected two billion people in 2019, representing 25% of the global population. Of these, close to 750 million people—10% of global population—suffer from severe food insecurity.

---

[10]Not necessarily a sequential continuum (see [6]: 1).

[11]"Development and (positive) peace actions also share the commitment to humanity and follow complementary principles (…). Humanitarian principles must be safeguarded (…) but humanitarians should also engage in conflict analysis, adopt conflict-sensitive programming, and collaborate with peace actors, where appropriate, to inform approaches which may ultimately contribute to peace outcomes" ([6]: 2). The IASC paper provides a full chapter of key considerations for humanitarians ([6]: 11f).

The outlook is all but encouraging: The State of Food Security and Nutrition in the World 2020 report states: "Projections show that the world is not on track to achieve Zero Hunger by 2030 and, despite some progress, most indicators are also not on track to meet global nutrition targets. The food security and nutritional status of the most vulnerable population groups is likely to deteriorate further due to the health and socio-economic impacts of the COVID-19 pandemic" [23].

Since conflict and insecurity are the main drivers of hunger and food crises and—in turn—hunger also drives conflict by fuelling long-standing grievances and disputes over land, livestock and other assets, WFP started collaborating with the Stockholm International Peace Research Institute (SIPRI) in 2018 to research the relationship between food and security (see [24]). The key pillars of the collaboration between SIPRI and WFP include

- Improving the evidence base for the relationship between food and security,
- Operationalizing findings in current and future programmes and
- Developing policies that clearly articulates WFP's contribution to peace and conflict prevention.

Country case studies under phase I of the cooperation indicate that WFP programming can and do make a positive contribution to improving the prospects for peace while there are also issues that need to be addressed.[12] The study concludes with 12 recommendations which including the following (see [24]: 30f):[13]

- Incorporate conflict analysis
- Build staff competence in conflict sensitivity
- Ensure a holistic programming approach
- Develop multi-sector, multi-stakeholder partnerships
- Generate contextually calibrated indicators of progress
- Include more qualitative methods in monitoring and evaluation
- Have a systematic approach to data visualization
- Familiarize with external data sources
- Supplement WFP skill sets and knowledge by working with partner organizations.

In the following concrete examples of WFP humanitarian assistance aiming at supporting food security will be presented and discussed.

Food assistance represents by far the most important activity of WFP in Palestine. The WFP Palestine Country Strategic Plan (2018–2022) plan focuses on two strategic outcomes:

1. Poor and severely food-insecure people have improved dietary diversity
2. Palestinian institutions have enhanced capacities and systems to identify, target and assist food-insecure vulnerable populations.

---

[12] Among the issues to be addressed are a missing theory of change which explains why a peace-positive outcome is expected and WFP staff not taking specifically and explicitly into account the conflict context ([24]: 2f).

[13] In the meantime, progress has been made in several regards. Inter alia WFP introduced Minimum Standards for Conflict Sensitive Programming which all WFP offices and HQ have to abide by [25].

Under strategic outcome 1, WFP Palestine implements the following activity: Provision of unconditional food assistance to currently 400.000 poor and food-insecure people by using the following modalities:

- Cash Based Transfers[14] (CBT): CBT e-vouchers are distributed via an online system, redeemable at any WFP-contracted shop.
- Food/in-kind: The ration covers 60% of the need and is composed of fortified wheat flour, pulses (lentil and chickpeas), fortified vegetable oil and iodized salt.

Complementing the provision of unconditional food assistance, tailored nutrition-sensitive awareness activities are implemented to engage diverse groups of men, women, girls and boys, raising their awareness of nutrition.

3.2   Development Work

Beside its humanitarian mandate WFP also has a development mandate. As set out in Article II of WFP's General Regulations the purposes of WFP are (see [22]):

- To use food aid to support economic and social development,
- To meet emergency and protracted relief food needs and
- To promote world food security in accordance with the recommendations of the United Nations and the Food and Agriculture Organization (FAO) of the United Nations.

The current Strategic Plan spells out as follows: "WFP's mandate allows it to apply development tools and perspectives to its humanitarian responses, providing communities with early recovery and development-enabling interventions that help build resilience and contribute to productive opportunities over the long term. As a result of this mandate, WFP's experience in both humanitarian and development contexts has allowed it to establish unique strengths and capacities to support food security and nutrition, including in contexts of protracted crises. In keeping with the 2030 Agenda, as WFP focuses on its core business of saving lives, it must do so in ways that contribute to outcomes that provide productive opportunities over the longer term, working collaboratively across institutional boundaries at the human-itarian—development and peace-building nexus, in line with the policy on WFP's role in peace-building in transition settings, while ensuring that it does not deviate from the primacy of humanitarian principles" [22].

In 2018, a dedicated WFP publication states: "WFP's mandate is executed through programmes that already transcend the humanitarian—development pillars (…). This makes WFP programmes well placed to not only contribute to both humanitarian and development objectives, but also to leverage its presence, tools and dual mandate to enhance its contribution to peace" ([18]: 5).

---

[14]The UN Agenda for Humanity calls to "use cash-based programming as the preferred and default method of support" ([13]: 9).

In Palestine, WFP started doing development work in 2019 [26, 27] focusing

- On increased (climate) resilience of its beneficiaries and the country at large by implementing a livelihood approach and
- On the modernization of the social assistance system by working together with ILO and UNICEF on the introduction of a Social Protection Floor.

In the (climate) resilience activity WFP started resilience activities which focus on providing climate-resilient agricultural assets to vulnerable households, particularly households headed by women, to improve their dietary intake and to improve their livelihoods. This is also based on the assumption, that "activities that improve the human capital and economic opportunities of the most vulnerable members of society by enhancing and or diversifying their livelihoods and improving their resilience may contribute to peacebuilding" [28]. According to KfW Development Research (see [29] infrastructure has the potential to contribute to peace and security—especially when it leads quickly and noticeably to an improvement in living conditions (which the provision of assets does).

In another contribution to a more sustainable work at the HDP nexus WFP is working with the government and other humanitarian actors, including UNICEF and ILO, to create an enhanced national Social Protection Floor (SPF) that is more accessible to the most vulnerable groups, particularly elderly people and persons with disabilities. This activity does not only reflect the principle of leaving no one behind (LNOB) but also takes into consideration that "when citizens are unable to access essential state services and social safety nets, they can feel neglected or marginalized (…). This (…) can spark unrest and violence, as notably witnessed in the context of food insecurity during the Arab Spring of 2011. In these situations, improving the equitable delivery of government services has been found to strengthen state-citizen relationships and to promote stability and the prospects for peace" [28]. The UN Agenda for Humanity also calls to "enhance national social protection systems that ensure equitable access to social services" ([13]: 9).

These development innovations are all about the HDP nexus: The number of poor and food insecure people (and in need of humanitarian assistance) keeps rising while WFP is not able to support all of them since funding to Palestine has come dramatically down. Among the 2,1 million poor and food insecure people there are many who are able and willing to work/engage in an activity, but due to the overall situation (occupation, the Covid-19 pandemic and its economic implications etc.) there are fewer and fewer jobs and people have no assets to start an activity on their own. Other aspects of the changing situation are the following:

- A financial crisis is limiting the government's ability to support the people in need (especially MoSD to provide cash transfers to poor people).
- Cash and food assistance are not (only) what people need. They also need to be and feel productive and live in dignity.
- Agriculture has gone down (3% of GDP) while the number of people in Palestine has gone up. Food dependency and food insecurity are growing.

- Climate change hits Palestine hard and will further worsen the situation unless climate smart agriculture will be introduced and used to a significant degree.
- The Covid-19 pandemic crisis has exacerbated all elements mentioned above.

Moreover, the WFP approach further developed in the meantime. WFP now acknowledges that "a broader mindset focused on the fostering of "people-owned" capacities (is required) in order to shift from WFP's perception of "delivering" resilience to people and communities" ([30]: 11).

Already in 2015 WFP stressed "the importance of a resilience-building approach" and stated: "WFP's practical experience across its humanitarian and development mandate offers some comparative advantage in enhancing resilience through food security and nutrition. (…) WFP will support resilience-building (…)" [31].

In 2017, WFP called for pro-smallholder food assistance "to overcome the simplistic perception of WFP's role in rural areas as a mere deliverer of unsustainable food handouts to passive recipients" [32].

In 2019, a strategic evaluation of WFP's support for enhanced resilience [30] acknowledged that "resilience building has long been implicit in WFP's work outside humanitarian settings. The evaluation identified commitment in WFP to contributing to resilience building (…). The more explicit treatment of resilience building in the WFP Country Strategic Plan (2017–2012) confirms this commitment" [30]. The evaluation explicitly recommended to "encourage country offices to adopt a resilience approach" ([30]: 14). The evaluation concluded that "WFP has the foundations for and a higher-level strategic commitment to enhancing resilience in order to ensure that individuals and communities can withstand shocks. This needs to be matched by and grounded in operational realities (…)" [30].

WFP Palestine therefore started the above-mentioned new activities and aims to include one or two new strategic outcome(s) in its next Country Strategic Plan:

- Improved household adaptation and resilience to climate and other shocks and/or
- Increased smallholder production and sales.

3.3 Human Security/Peace

People from very different walks of life have acknowledged the importance of food with regard to peace. The German playwriter and poet Bertold Brecht stated in "The Threepenny Opera": "Erst kommt das Fressen, dann kommt die Moral" (1928).[15] The German politician Willy Brandt said in his speech to the UN General Assembly in New York in 1973: „Wo Hunger herrscht, ist auf die Dauer kein Friede."[16]

If conflict is the most important reason for hunger (80% of humanitarian needs arise from violent conflict, 60% of the world's hungry live in conflict areas; see [33]

---

[15]Food comes first, then morals.

[16]Where mass hunger reigns, we cannot speak of peace.

and the single greatest challenge for achieving zero hunger, it is of utmost importance to tackle food insecurity in order to achieve human security[17] and peace.

But since development cooperation discovered conflict management, peace-building and crisis prevention as an area of relevance and interest in the 1990s (see [35]) little progress has been made in achieving peace and security (see Sect. 2). This is not the fault of development cooperation. There are just too many more conflicts and crises. The concepts and approaches developed since then may also not necessarily be the most important ones.

Food security however is of fundamental importance to prevent the escalation of violence. Food assistance is not only a lifesaver for people trapped in conflict, living under siege or on the run after being forced out of their homes. It is also the first step towards peace by helping to ease tensions that could escalate into conflicts or restart violence. This is not only the view of the WFP. It was confirmed by research: "Recent studies (...) suggest that hunger, especially in the form of rising food prices and competition for resources, can exacerbate tensions and contribute to conflict" ([11]: 2). And also the Nobel Peace Prize recognizes WFP as a peace maker, concretely "for its efforts to combat hunger, for its contribution to bettering conditions for peace in conflict-affected areas and for acting as a driving force in efforts to prevent the use of hunger as a weapon of war or conflict" (see [8]).

4   The Specificities of the Country Case of Palestine

Section 3 has shown concretely that work at the nexus can and should be done. Howe ([11]: 11) stresses that "it is not so much a question of whether to use a triple nexus approach, but rather how it should be adapted to the particular context in order to harness, to the extent possible, the potential impact of actions. (...) By taking a clear-sighted approach and adjusting for different and evolving contexts, it (the triple nexus framework, HL) has the potential to contribute to the achievement of the Sustainable Development Goals, especially in countries affected by conflict".

Weishaupt [36] also points to the importance of "contextual particularities, as they determine the specific needs for humanitarian aid, development and peace as well as the potential for meeting them in a comprehensive/collective manner".

While the previous section looked into some country examples "from the field", all from Palestine, this section explains the specificities and limitations of the country case.

---

[17]The Commission for Human Security (CHS), in its final report Human Security Now, defines human security as: "... to protect the vital core of all human lives in ways that enhance human freedoms and human fulfilment. Human security means protecting fundamental freedoms—freedoms that are the essence of life. It means protecting people from critical (severe) and pervasive (widespread) threats and situations. It means using processes that build on people's strengths and aspirations. It means creating political, social, environmental, economic, military and cultural systems that together give people the building blocks of survival, livelihood and dignity" ([34]: 4).

Palestine is different from other countries as it is under occupation and therefore suffering from limited control over the country[18] plus ongoing and expanding settlement activities which limit their possibilities to reduce or end needs. The UN General Assembly (in several resolutions including resolution 74/243) demanded that Israel, the occupying Power, cease the exploitation, damage, cause of loss or depletion and endangerment of the natural resources in the Occupied Palestinian Territory. As a result, "the long-term economic repercussions of practices, policies and measures applied by Israel have entailed low investments in the Palestinian economy, which has led to a process of de-industrialization, the erosion of the Palestinian productive sectors and to de-development, specifically in Gaza. This has entrenched the dependency of the Palestinian economy on Israel and on foreign aid" [38].

The following is mainly based on the latest UN report on the situation (it was the UNSG note to the UN Economic and Social Council of the UN General Assembly about the economic and social repercussions of the Israeli occupation on the living conditions of the Palestinian people, see [38]. The range of issues Palestine suffers under occupation include ongoing settlement activities, discriminatory allocation of land and construction permits, discriminatory provision of services, no equality before the law, violence and use of force, displacement of populations, destruction and confiscation of property and infrastructure, movement and access restrictions including the closure of Gaza, exploitation, endangerment and depletion of Palestinian natural resources and limited opportunities for agricultural production:

Settlement activities: In resolution 73/255 (2019), the UN General Assembly stressed that the wall and the settlements being constructed by Israel in the Occupied Palestinian Territory are contrary to international law. This has been confirmed in other resolutions, such as UN Security Council resolution 2334 (2016), in which the Security Council "reaffirmed that the establishment by Israel of settlements in the Palestinian territory occupied since 1967, including East Jerusalem, had no legal validity and constituted a flagrant violation under international law and a major obstacle to the achievement of a lasting and comprehensive peace. The establishment and expansion of Israeli settlements in the Occupied Palestinian Territory amounts to the transfer by Israel of its own civilian population into the territory it occupies, which is prohibited under international humanitarian law" [38].

Discriminatory allocation of land and construction permits: "The Israeli zoning and planning policies in Area C, which constitutes 60% of the West Bank, and East Jerusalem, are discriminatory and are considered incompatible with requirements under international law" [38]. Land is allocated "almost exclusively to Israeli settlements or to the military and facilitating the growth of Israeli settlements (…)" [38].

---

[18]"The Paris Protocol entrenched the dependence of the Palestinian economy on Israel via customs union that leaves no space for independent Palestinian economic policies. It ties the Occupied Palestinian Territory to the trade policies, tariff structure and value-added tax rate of Israel. Moreover, the authorities in Israel collect trade tax revenues on behalf of the Palestinian National Authority and transfer them to PNA. This arrangement allows Israel to control two thirds of Palestinian tax revenue, a leverage frequently used…" ([37]: 5).

Discriminatory provision of services: "Israeli policies and practices also entail discrimination in service provision. For example, Palestinians constitute 30% of the population in Jerusalem. They pay 40% of the total value in taxes that the Israeli municipality collects, yet, the municipality only allocates 8% of tax revenues to the services provided to the Palestinians" [38].

No equality before the law: "Palestinians in the occupied territory continue to be subject to a complex combination of Israeli and Palestinian legal systems. In the West Bank, Israeli domestic law is applied extraterritorially to Israeli settlers, while Palestinians are subject to Israeli military law in addition to the Palestinian legal system. (…) The application of two different legal systems in the same territory on the sole basis of nationality or origin is inherently discriminatory and violates the principle of equality before the law, which is central to the right to a fair trial" [38].

Violence and the use of force: "As the occupying Power, Israel has the obligation to take all the measures in its power to restore and ensure, as far as possible, public order and life in the Occupied Palestinian Territory and to protect the Palestinian population from all acts of violence, in all circumstances (see A/74/357, para. 26). Practices of the military and security forces of Israel continue to raise concerns, especially with respect to the excessive use of force and, in some cases, unwarranted force amounting to arbitrary deprivation of life" [38].

Population displacement: "There is continued concern that a combination of Israeli policies and practices in Area C, East Jerusalem and the city of Hebron (…) have created a coercive environment. Involuntary displacement and relocation to alternative residential areas as a result of such policies may amount to forcible transfer if it is carried out without the free and informed consent of the individuals who relocate, in violation of the obligations of Israel under international humanitarian and human rights law" [38].

Destruction and confiscation of property and infrastructure: "Moreover, since the onset of the occupation, Israel has completely demolished around 50,000 residential units and partially destroyed more than 100,000" [38].

Movement and access restrictions: "As freedom of movement is a prerequisite to the exercise of other human rights, such as the rights to family, health and education, the closures and related practices imposed by the Israeli authorities, in particular the restrictions on movement, have had a devastating impact on the lives of Palestinians, in particular on families (…). The restrictions have fragmented the Palestinian landscape, including the separation of Gaza from the West Bank. They have created isolated communities, undermined social cohesion, ruptured a common identity and reduced economic activity within and among the fractured Palestinian population" [38]. Assessing the costs of mobility restrictions, the World Bank estimates that relaxation of restrictions by Israel could enlarge the Palestinian economy by 33%" ([37]: 19).

Gaza closure: "Imposed since June 2007, following the takeover by Hamas, the closures in Gaza, which affect the movement of goods and people, continue to undermine the civil, political, economic, social and cultural rights of Palestinians in Gaza and continue to affect all areas of Palestinian life. The blockade may amount to collective punishment, which is prohibited under international law" [38]. In economic terms the impact of the closure "entrenched the dependence of more than 80% of the population on international assistance" ([37]: 21).

Exploitation, endangerment and depletion of Palestinian natural resources: "Area C continues to be almost entirely off limits for the Government of Palestine, as well as for producers and investors, even though it contains the most valuable natural resources. (…) Closures, particularly in Gaza, limit Palestinian access to materials and technologies. Approximately 3.7 million Palestinians are negatively affected by a lack of access to safe water, sanitation and hygiene services. Almost the whole population in Gaza is exposed to public health risks associated with poor water quality (…)" [38] while Israel controls 85% of Palestinian water sources" ([37]: 25).

Limited opportunities for agricultural production: "The total area of land classified as being of high or medium agricultural value in the West Bank is 2,072,000 dunums, which constitutes about 37% of the West Bank. Palestinians can only use less than half of that area, mainly owing to land confiscation and the restriction of Palestinian farmers' access to water and land. (…) Furthermore, they have to buy either water from Israel or desalinated water from private suppliers at a high cost, thereby reducing their competitiveness in the market. The agricultural potential of Gaza has been undermined by the closure, as some 35% of farmland falls within restricted areas enforced by Israel. Furthermore, Israel has damaged Palestinian farmland in Gaza by aerially spraying the land with herbicides" [38].

Food insecurity and the risk of heightened levels of violence: All these practices directly impact on the food insecurity. Food insecurity in Palestine is driven by high levels of unemployment (33% in 2019, see [37] and poverty (29.2% in 2019, see [37] which are linked to occupation including the blockade of Gaza (see also [38]). Prior to the outbreak of the Covid-19 pandemic, food insecurity in Palestine affected one third of the population (1.7 million people). In Gaza, almost three quarters of the population were food insecure. Under the impact of Covid-19 the already very precarious situation (especially in Gaza) will further deteriorate. Rising unemployment and poverty will lead to higher numbers of (severely) food insecure people. According to the 2020 UN Humanitarian Needs Overview, more than 2 million people are now in need of food assistance. This number represents a net increase of approximately 300,000 newly food-insecure people. The worsening of socio-economic and food insecurity indicators have also been confirmed by the Palestinian Central Bureau of Statistics (PCBS): A PCBS survey done recently (end of 2020) reveals that the income of 42% of families decreased by 50% and more. To cope with this, 41% of the families decreased their monthly spending on food.

In 2020, WFP provided food assistance to more than 400.00 of the poorest and most food insecure non-refugees[19] in the Gaza Strip, the West Bank and East Jerusalem. Food assistance was predominantly provided through cash-based transfers (it was electronic food vouchers which allow beneficiaries to buy food in accredited shops), but also through in-kind food parcels. The impact of cash-based transfers goes beyond improving the dietary intake of food insecure families. Cash based transfers also have a positive spillover effect on the local economy (producers, retailers, shops

---

[19]Refugees are taken care of by a dedicated UN agency, the UN Relief and Work Agency (UNRWA).

etc.). WFP's cash-based transfer system can be and is used by various other actors such as sister UN agencies, INGOs and NGOs.

To be able to continue supporting food insecure people in Palestine and by doing so contributing to peace, WFP needs to secure funding: reliable regular funding[20] is critical not only to alleviate the deteriorating food security and the worst effects of the socio-economic consequences of the pandemic, but also to act as a stabilizing factor in an uncertain political and socio-economic climate.

The UN had already projected years ago that by 2020, Gaza would become an unlivable place [39]. In addition to the health implications of the pandemic, the loss of jobs and incomes "will have profound socioeconomic implications. Coupled with the effects of the protracted occupation and the matrix of Israeli policies and practices, the social outlook in the Occupied Palestinian Territory seems bleak" [38]. All people, but especially young people, suffer from psychological effects of the worsening situation. The UNSG is worried about "the risk of social unrest. Youth are the most detrimentally affected and experience very high unemployment rates. They may undergo a psychological impact that later contributes to heightened levels of violence" [38].[21]

To conclude, Palestine is not a typical case of a country suffering a humanitarian crisis. The humanitarian needs and development challenges are mainly related to occupation and the political situation. It is a protracted political crisis which made Palestine one of the top ten recipients of humanitarian and development assistance for many years. In times of shrinking humanitarian and ODA budgets on the one hand[22] and increased needs all over the world (last but not least because of the Covid-19 pandemic) Palestine has no choice but to adapt, it was to achieve more with less. A nexus approach seems to be the perfect answer to this challenge as it has the potential to stop duplication, to save time and resources and to improve targeting and implementation.

Despite (or precisely because of) the special nature of the case, Palestine needs all attention and support. Any further worsening of the situation may lead to an escalation

---

[20]WFP is funded by (relatively unpredictable) voluntary contributions only.

[21]"Half of the population is ready for violence—even more so among younger people" [40]. The World Bank and UN study "Pathways for peace" stressed the importance of this: "Some of the greatest risks of violence today stem from the mobilization of perceptions of exclusion and injustice, rooted in inequalities across groups. When an aggrieved group assigns blame to others or to the state for its perceived economic, political, or social exclusion, then emotions, collective memories, frustration over unmet expectations, and a narrative that roused a group to violence can all play a role in mobilization to violence" ([5]: xxii).

[22]"Donor budget support has declined substantially in recent years, falling from 32% of GDP in 2008 to 3.5% of GDP in 2019. The negative trend in aid, combined with unpredictability and fluctuations, has been a constant source of fiscal uncertainty" ([37]: 9). On top of this comes the decline in funding to UN agencies operating in Palestine. This "aggravates fiscal stress by constraining economic growth on the demand side and increasing the pressure for transfers to poor households. The social and humanitarian spending of these agencies stimulates the economy and provides jobs and critical services" ([37]: 9).

of violence.[23] The UN Secretary General therefore called for both—attention and support: "The unprecedented challenges posed by the COVID-19 crisis heightens the vulnerability of Palestinians, in particular the population of Gaza, and exposes them to more risk. Palestine refugees and youth, whose social and economic conditions were already precarious, stand to suffer disproportionately both from the pandemic and its aftermath." [38].

No matter what, WFP and the UN country team "will continue to ensure that the United Nations works towards the establishment of an independent, democratic, contiguous and viable Palestinian State, living side by side in peace with a secure Israel, with Jerusalem as the capital of both States, consistent with relevant Security Council resolutions and international law" [38]. But there are however (at least) two main pushbacks that need to be overcome:

1. The Palestinian Government is reluctant to move from humanitarian support to more sustainable approaches as they tend to see assistance as an entitlement which they do not want to give up on. This needs to be understood in the context of the long history of injustice that was done to Palestine. It will not be overcome before Palestine will be recognized as an independent state (which is long overdue). But humanitarian and development assistance cannot compensate for overdue political solutions. Both leave the primary responsibility for the wellbeing of the country and its population with the Government.
2. Most actors got accustomed to working in the comfort zone of their silos. The idea of working at the triple HDP nexus is not necessarily well received by all actors (international organisations, bilateral development agencies, international and national agencies …) working in Palestine. There are mainly two actors who can and should take action: the Government of Palestine and the Resident Coordinator.[24] It is to be hoped that they see the importance of improved joint work at the HDP nexus and that they will act accordingly.

5 Conclusions and Recommendations

Despite little progress over the five years since the triple HDP nexus concept emerged it is still relevant and worth pursuing. Its "real strength is that it creates multiple, mutually reinforcing impact pathways. (…). First, there is a magnified direct impact. (…) Second, a nexus action contributes to other SDG outcomes. (…) Third, these contributions to other SDGs have an indirect reinforcing impact on the targeted collective outcome of SDG 2. (…) Forth, the direct and indirect impacts on SDG 2, in turn, contribute to wider peace, development and humanitarian efforts" ([11]: 9). The talk needs to be walked now for countries to effectively benefit from the famous

---

[23] According to Kubovich, even senior Israeli defense officials "said that these (it was Covid-19 related, HL) changes are bringing Palestinians society closer to the boiling point" [41].

[24] The UNSG's Agenda for Humanity clearly calls for "the resident/ humanitarian coordinator to ensure coherent, collective and predictable programme delivery by the United Nations and its partners towards (…) the achievement of collective outcomes" ([13]: 11).

triple HDP nexus concept. Conclusions and recommendations to make this happen include the following:

- "Unzureichende Hilfe fuehrt zu wachsender globaler Instabilitaet" ([42]: 76). Therefore, there is a dire need to reverse the downward trend of humanitarian and development funding in order to make sure that Human Response Plans and the UN Central Emergency Response Fund (CERF) are fully funded.[25] China and the Arab countries for example contribute very little to the financing of multilateral humanitarian aid (see [42]). A solidarity levy as called for in the run-up to the World Humanitarian Summit 2016 in Istanbul or innovative financing mechanisms such as climate risk insurance solutions would be other ideas to increase funding.
- Reorganize the financing of global humanitarian assistance (see [42]: 75). It should be seen and organized as global public obligation towards a global public good (instead of a voluntary gesture). The Agenda for Humanity called for "assistance frameworks over 10 to 15 years" ([13]: 3).
- Provide more multilateral than bilateral aid (currently 77% of all ODA is channelled through bilateral structures) to incentivize and support common approaches and outcomes.
- Move away from "a 'one-size-fits-all' model to differentiated nexus configurations" [36] which need to be crafted, drafted and implemented on the ground. Evidence has shown that work at the nexus "was only possible when (...) approached as bottom-up processes, rooted in local contexts" [43].
- Support governments to design and implement inclusive social protection systems and floors. This may start with humanitarian assistance, can be further developed and integrated into a government owned social protection system and promote peace by avoiding people to turn to extremism just because they do not have anything to live on.
- Implement the reform of the UN development system: A "One UN" approach instead of the existing multitude of specialised organisations competing with each other will be much more effective as it will allow for transformative approaches instead of patchwork.
- To achieve this, "a sensitivity towards the nexus—that is towards interrelations and interdependencies between humanitarian aid, development and peace" needs to be cultivated within UN organizations" [36].

Countries around the world—not only, but also Palestine—would benefit a lot from any progress made.

---

[25]The UN Agenda for Humanity ([13]: 14) called for the following: "Increase the coverage of inter-agency humanitarian appeals to a minimum average of 75 percent per year by 2018." "Expand the Central Emergency Response Fund from USD 500 million to USD 1 billion by 2018." This is fraction of the cost of humanitarian assistance which may reach "a staggering US$ 50 billion per year" in 2030 (World Bank 208: 1). But humanitarian funding is going down since 2017 and so does official development assistance since 2018 (see [43]).

For Palestine, "a fully re-engaged peace process that leads to an end of the occupation and a realization of a two-State solution is the best and perhaps only vehicle for ultimately addressing humanitarian needs, accelerating development, and building peace" ([44]: 4).

References

1. UN (2020) Progress towards the sustainable development goals, report of the secretary-general. New York. https://undocs.org/en/E/2020/57
2. SIPRI Yearbook (2020) Armaments, disarmament and international security. SIPRI, Stockholm
3. OECD (2020) States of fragility report 2020. Paris
4. Senghaas D (1988) Konfliktformationen im internationalen system. Frankfurt am Main
5. World Bank (2018) Pathways for peace, inclusive approaches to preventing violent conflict. Washington D.C, World Bank
6. IASC (2020) Exploring peace with the humanitarian—development—peace nexus (HDPN). Geneva 2020
7. Sen (1999) Armatya: democracy as freedom
8. Norwegian Nobel Committee (2020) Nobel Peace Prize laudatio, Oslo, 9 Oct 2020
9. International Dialogue (2011) On peacebuilding and statebuilding: new deal for engagement in Fragile States. Busan
10. UN (2015) 2030 Agenda for sustainable development. New York
11. Howe P (2019) The triple nexus: a potential approach to supporting the achievement of the Sustainable Development Goals? World Dev 124(219):104629
12. Stockholm Declaration (2016) On addressing fragility and building peace in a changing world. Stockholm
13. UN Secretary General (2016) Agenda for humanity, Annex to the Report of the SG for the World Humanitarian Summit, A/70/709, 2. Feb 2016
14. OCHA (2017) New way of working, Geneva
15. UN (2016) Secretary-General-designate António Guterres' remarks to the General Assembly on taking the oath of office. New York. https://www.un.org/sg/en/content/sg/speeches/2016-12-12/secretary-general-designate-ant%C3%B3nio-guterres-oath-office-speech
16. UN (2017) Restructuring of the United Nations peace and security pillar: Report of the Secretary-General, United Nations General Assembly. New York. https://undocs.org/en/A/72/525
17. OECD/DAC (2020) DAC recommendations on the humanitarian-development-peace nexus. OECD, Paris
18. World Food Programme (2018a) WFP and the humanitarian, development and peace nexus. WFP, Rome 2018
19. Lingnau H et al. (2019) Humanitarian—Development—Peace Nexus in Palestine: Nexus options and a way forward, concept note, Dec 2019
20. OCHA (2019) Operationalizing collective outcomes—Lessons learned and best practices from and for country offices. Geneva
21. Pantuliano (2018) Sara: delivering the UNSG's prevention vision. ODI, London. https://www.odi.org/blogs/10623-pathways-peace-our-experts-impressions-new-report
22. World food programme (2016) WFP strategic plan 2017–2021. Rome
23. FAO (2020) The state of food security and nutrition in the world 2020, Rome
24. Delgado C, Jang S, Milante G, Smith D (2019) The world food programme's contribution to improving the prospects for peace, Preliminary report. SIPRI, Stockholm June 2019
25. World Food Programme (2020) WFP minimum standards for conflict sensitive programming. Rome

26. Lingnau H (2019) Empowering WFP beneficiaries to increase their resilience via different livelihood activities, concept note, July 2019
27. Lingnau H (2020) Why WFP COP is considering adding a 3rd SO (and respective activity) on resilience to its Country Strategic Plan, concept note, May 2020
28. World Food Programme (2019a) Triple Nexus: WFP's contributions to peace. WFP, Rome
29. Prigge-Musiat J (2020) Friedensfoerderung durch Infrastruktur? KfW, Frankfurt am Main
30. World Food Programme (2019b) Summary report on the strategic evaluation of WFP's support for enhanced resilience. Rome
31. World Food Programme (2015) Policy on building resilience for food security and nutrition. Rome
32. World Food Programme (2017) Pro-smallholder food assistance—a strategy for boosting smallholder resilience and market access worldwide. Rome
33. World Food Programme (2018b) Fact sheet Hunger and conflict. Rome 2018
34. Commission on Human Security (2003) Human security now: final report. CHS, New York
35. Grossmann G, Lingnau H (2002) Vergangenheits- und Versöhnungsarbeit - wie die TZ die Aufarbeitung von gewaltsamen Konflikten unterstützen kann, Eschborn
36. Weishaupt S (2020) The Humanitarian-development-peace nexus: towards differentiated configurations. UNRISD Working Paper 2020-8
37. UNCTAD (2020) Report on UNCTAD assistance to the Palestinian people: developments in the economy of the occupied Palestinian Territory. UNCTAD, Geneva 5 Aug 2020
38. UN Secretary General (2020) Note to the UN economic and social council of the UN general assembly: economic and social repercussions of the Israeli occupation on the living conditions of the Palestinian people in the Occupied Palestinian Territory, including East Jerusalem, and of the Arab population in the occupied Syrian Golan (27.5.2020) A/75/86–E/2020/62
39. UN (2012) Gaza in 2020: a liveable place? New York
40. Pfeffer A (2020) Twenty years after second intifada, is a third likely? Haaretz 23 Oct 2020
41. Kubovich Y (2020) Security brass fear PA financial crisis may spur violence, The pandemic has driven unemployment to 35% in what officials call the worst situation in the PA in a decade. Haaretz 29 Sept 2020
42. Engelhardt M (2018) Weltgemeinschaft am Abgrund—Warum wir eine starke UNO brauchen, Bonn
43. Caparini M, Reagan A (2019) Connecting the dots on the triple nexus, SIPRI commentary Stockholm 29.11.2019
44. Office of the UN Special Coordinator for the Middle East Peace Process (2020) Socioeconomic report

Location Intelligence Powered by Machine Learning Automation for Mapping Malaria Mosquito Habitats Employing an Unmanned Aerial Vehicle (UAV) for Implementing "Seek and Destroy" for Commercial Roadside Ditch Foci and Real Time Larviciding Rock Pit Quarry Habitats in Peri-Domestic Agro-Pastureland Ecosystems in Northern Uganda

Benjamin G. Jacob and Peace Habomugisha

Abstract Public health emergencies stemming from infectious disease outbreaks is creating a serious threat to global health security. For example, climate change and extreme weather events threaten to alter and affect geographic areas pertaining to disease vulnerability, such as greater risks of mosquito-borne diseases (dengue, malaria, yellow fever and Zika). The emergence of these disease outbreaks and their influence globally has sparked a renewed attention on global health security and the application of location intelligence. Persistent outbreaks characterize a 'new normal' that points to major deficiencies in preparedness, response and recovery initiatives. Malaria mosquito *An. gambiae s.l., arabiensis s.s.* and *funestus s.s* represent the main malaria mosquito vectors in sub-Saharan Africa. As reported in WHO (Jacob et al. in Open Remote Sensing 17:11–24, [1]), Malaria is a life-threatening disease caused by parasites that are transmitted to people through the bites of infected female Anopheles mosquitoes. It is preventable and curable. In 2019, there were an estimated 229 million cases of malaria worldwide. The estimated number of malaria deaths stood at 409,000 in 2019. Children aged under 5 years are the most vulnerable group affected by malaria; in 2019, they accounted for 67% (274,000) of all malaria deaths worldwide. The WHO African Region carries a disproportionately high share of the global malaria burden. In 2019, the region was home to 94% of malaria cases and deaths. Sensemaking lies at the heart of location intelligence. Location intelligence is defined as the collection and analysis of geospatial data that are

B. G. Jacob (✉)
College of Public Health, University of South Florida, Tampa, FL, USA
e-mail: bjacob1@usf.edu

P. Habomugisha
Carter Center, Kampala, Uganda

transformed into strategic insights to support operations. Weick (Krizhevsky et al. in Advances in Neural Information Processing Systems, pp 1097–1105, [2]) refers to sensemaking in terms of '…how we structure the unknown so as to be able to act in it. Sensemaking involves coming up with a plausible understanding—a map—of a shifting world; testing this map with others through data collection, action, and conversation; and then refining, or abandoning, the map depending on how credible it is' (Lin et al. in Proceedings of the IEEE International Conference on Computer Vision 2017, pp. 2980–2988, [3]). The application of machine learning algorithms are emerging as key public health intelligence approaches to support tactical, operational and strategic sensemaking. Recent advances that identify the reflective signatures of active mosquito breeding sites, and their temporal evolution, have made predictive algorithms possible to search and identify previously unidentified larval habitats from a Unmanned Aerial Vehicle (UAV), and monitor their activity in real time. Spectral signature is the variation of reflectance of a material (i.e., emittance as a function of wavelength) (www.esri.com). These real time aerial surveys can provide spatiotemporal data for targeting interventions to eliminate vectors before they become adult airborne biting mosquitoes, to reduce malaria transmission. Reference capture point habitats for *Anopheles gambiae s.l., An. arabiensis s.s.* and *An. funestus s.s,* the main malaria mosquito vectors in sub- Saharan Africa [www.who.int], may also be separately identified with this methodology. This chapter points to the application of predictive algorithms coupled with drone surveillance to support sensemaking in support of spatiotemporal data for targeting interventions to eliminate vectors before they become adult airborne biting mosquitoes, to reduce malaria transmission. The sensemaking applies not only to the targeted interventions to eliminate vectors, but also strategic sensemaking that contextualizes this intervention as part of a more holistic/systemic and strategic intervention encompassing a myriad of coordinated interventions across the disaster management spectrum (mitigation, preparedness, response, recovery).

Keywords Malaria · Predictive algorithm · Sensemaking · Capacity building

1 Introduction

Public health emergencies stemming from infectious disease outbreaks is creating a serious threat to global health security. For example, climate change and extreme weather events threaten to alter and affect geographic areas pertaining to disease vulnerability, such as greater risks of mosquito-borne diseases (dengue, malaria, yellow fever and Zika). The emergence of these disease outbreaks and their influence globally has sparked a renewed attention on global health security and the application of location intelligence.

Persistent outbreaks characterize a 'new normal' that points to major deficiencies in preparedness, response and recovery initiatives. Malaria mosquito *An. gambiae s.l., arabiensis s.s.* and *funestus s.s* represent the main malaria mosquito vectors in sub-Saharan Africa. As reported in WHO [4], Malaria is a life-threatening disease caused by parasites that are transmitted to people through the bites of infected female Anopheles mosquitoes. It is preventable and curable. In 2019, there were an estimated 229 million cases of malaria worldwide. The estimated number of malaria deaths stood at 409,000 in 2019. Children aged under 5 years are the most vulnerable group affected by malaria; in 2019, they accounted for 67% (274,000) of all malaria deaths worldwide. The WHO African Region carries a disproportionately high share of the global malaria burden. In 2019, the region was home to 94% of malaria cases and deaths.

Sensemaking lies at the heart of location intelligence. Location intelligence is defined as the collection and analysis of geospatial data that are transformed into strategic insights to support operations. Weick [5] refers to sensemaking in terms of '…how we structure the unknown so as to be able to act in it. Sensemaking involves coming up with a plausible understanding—a map—of a shifting world; testing this map with others through data collection, action, and conversation; and then refining, or abandoning, the map depending on how credible it is' [6].

The application of machine learning algorithms are emerging as key public health intelligence approaches to support tactical, operational and strategic sensemaking. Recent advances that identify the reflective signatures of active mosquito breeding sites, and their temporal evolution, have made predictive algorithms possible to search and identify previously unidentified larval habitats from a Unmanned Aerial Vehicle (UAV), and monitor their activity in real time. Spectral signature is the variation of reflectance of a material (i.e., emittance as a function of wavelength) (www.esri.com). These real time aerial surveys can provide spatiotemporal data for targeting interventions to eliminate vectors before they become adult airborne biting mosquitoes, to reduce malaria transmission. Reference capture point habitats for *Anopheles gambiae s.l., An. arabiensis s.s.* and *An. funestus s.s,* the main malaria mosquito vectors in sub- Saharan Africa [www.who.int], may also be separately identified with this methodology.

This chapter points to the application of predictive algorithms coupled with drone surveillance to support sensemaking in support of spatiotemporal data for targeting interventions to eliminate vectors before they become adult airborne biting mosquitoes, to reduce malaria transmission.

The sensemaking applies not only to the targeted interventions to eliminate vectors, but also strategic sensemaking that contextualizes this intervention as part of a more holistic/systemic and strategic intervention encompassing a myriad of coordinated interventions across the disaster management spectrum (mitigation, preparedness, response, recovery).

As described in Crayne and Medeiros [7], 'Sensemaking is broadly defined as the process by which individuals interpret cues within a changing environment and use that interpretation to explain what has occurred and to promote future action'. In support of actionable public health intelligence, sensemaking leverages a '…backward-facing process, in which one collects information about a situation or event and attempts to develop an explanatory narrative (i.e., "What's the story?") that is then used as the basis for decisions and action [8] (cited in [7]). Herein lays the application of predictive analytics to support decision making and intervention strategies.

Sensemaking (through the application of location intelligence) becomes a game changer for threat, risk and vulnerability analysis. The tactical, operational and strategic interventions derived can be scaled up to support capacity building with long term impacts on community sustainability and resilience.

2 Operationalizing 'Seek and Destroy' Malaria Habitat Strategy

Recent advances that identify the reflective signatures of active mosquito breeding sites, and their temporal evolution, have made predictive algorithms possible to search and identify previously unidentified larval habitats from a Unmanned Aerial Vehicle (UAV), and monitor their activity in real time. Spectral signature is the variation of reflectance of a material (i.e., emittance as a function of wavelength) (www.esri.com). These real time aerial surveys can provide spatiotemporal data for targeting interventions to eliminate vectors before they become adult airborne biting mosquitoes, to reduce malaria transmission. Reference capture point habitats for *Anopheles gambiae s.l., An. arabiensis s.s.* and *An. funestus s.s,* the main malaria mosquito vectors in sub-Saharan Africa [www.who.int], may also be separately identified with this methodology.

An aerial indexed, capture point, time series, temporal signature framework represents a vital component for retrieval systems where a user submits an query video and a real time system retrieves a ranked list of visually similar land use land cover (LULC) habitat types (e.g., a hyperproductive, seasonal, *An. gambiae,* roadside ditch) by differentially corrected GPS coordinates which has a positional accuracy of 0.178 m (see [9]). The sensitivity and specificity of the video analog signals at identifying multiple, grid-stratifiable, LULC, eco-georeferenceable, larval habitats capture points can be subsequently evaluated by real time seasonal identification of the *Anopheles* breeding sites in natural settings (e.g., peri-urban, agricultural fields during pre-rain sample frames), followed by field verification (i.e., "ground truthing") of predictively mapped larval habitats. For remote identification of vector habitat, time series signatures the first step is often to construct a discrete tessellation of the region [10]. This data may then subsequently be fed into artificial intelligence (AI) algorithms employing a real time app with associated software created to find other habitats with similar characteristics from new surveys of unknown terrain.

In computer science, AI, (i.e., machine intelligence), is intelligence demonstrated by machines, in contrast to the natural intelligence displayed by humans and animals. Leading AI textbooks define the field as the study of "intelligent agents": any device that perceives its environment and takes actions that maximize its chance of successfully achieving its goals [11]. Colloquially, the term "artificial intelligence" is often used to describe machines (or computers) that mimic "cognitive" functions that humans associate with the human mind, such as "learning" and "problem solving".

As machines become increasingly capable, tasks considered to require "intelligence" are often removed from the definition of AI, a phenomenon known as the AI effect. A quip in Tesler's Theorem says "AI is whatever hasn't been done yet."[11]. For instance, optical character recognition is frequently excluded from things considered to be AI, having become a routine technology. Many tools are used in AI, including versions of search and mathematical optimization, artificial neural networks, and methods based on statistics, probability and economics. The AI field draws upon computer science, information engineering, mathematics, psychology, linguistics, philosophy, and many other fields. The field was founded on the assumption that human intelligence "can be so precisely described that a machine can be made to simulate it".

Here we employ a proposed computer vision approach for real time LULC mapping malaria mosquito *An. gambiae s.l., arabiensis s.s.* and *funestus s.s* larval habitats which was based on a Faster Region-based Convolutional Neural Network (Faster R-CNN) algorithm employing seasonally retrieved video analog datasets in a UAV real time platform. Faster R-CNN algorithm is a state of the art AI technique that not only classify entities within an image, but can also localize where the entities of interest are within an image [12].

In our proposed technique, we first learn the core features of the breeding site capture point LULC objects in the seasonal sampled UAV imaged datasets using convolutional neural networks (CNNs), and then integrate manual ground truthing (i.e., field verification) of identified larval habitats and their eco-georeferenceable geolocations within the images, with learned feature maps from prior training in order to design a final region-based neural network for classification and localization. Subsequently at run-time, we assumed our network could classify unseen images as larval habitats.

We assumed a text-based, real time, retrieval system may be implementable in the future (e.g., where a user submits a textual description of an eco-georeferenced, *Anopheles*, aquatic, breeding site, capture point) and the real time portal retrieves a ranked list of relevant unknown habitat, capture point, gridded, LULC locations based on GPS ground coordinates. This real time portal we assumed could facilitate this retrieval by finding the video clip of the unknown larval habitats that were manually assigned similar textual description (annotations) which may subsequently be employed for enabling the video anlaog signature framework through an iOS or Android application (app).

The overall goal of this project was to develop a customized smartphone app that could identify the eco- georeferenceable, LULC geolocation of unknown, *Anopheles,(gambiae s.l., funestus s.s., arabiensis s.s.)* capture point, agro-village, pastureland habitats from AI processed real time video images employing time series signatures obtained from a drone aircraft in Akonyibedo village in Gulu District, Northern Uganda. We assumed that once an anopheline capture point, LULC, larval habitat site, aquatic breeding site foci was identified and field validated, local village and entomological teams could be mobilized to the mapped habitat location employing the app and the GPS coordinates of the mapped site. Subsequently we assumed that an environmentally friendly tactic ("Seek and Destroy") could be used to bury the smaller habitats along households using soil substrate while larger habitats may be treated with a drone spray application using an environmental friendly insecticide (Abate).

The objectives of the research described in this article was: (1) to design, deploy and validate AI algorithms operating on drone videos (2) develop robust formulae for computing the swath, capture point, LULC coverage for generating field maps (3) construct a smartphone application (app) in order to enable automatic detection of potential larval habitats from drone videos for predictive mapping unknown unsampled geolocations of potential, seasonal, malaria, mosquito vectors, *An. gambiae s.l., arabiensis* s.s. and *funestus* s.s., sub-county, capture point, larval habitats in an agro-village, epi-entomological, intervention site (Akonyibedo village) in Gulu District, Northern Uganda.

3   Method

3.1   Study Site

Uganda lies between the eastern and western sections of Africa's Great Rift Valley. The country shares borders with Sudan to the north, Kenya to the east, Lake Victoria to the southeast, Tanzania and Rwanda to the south and the Democratic Republic of Congo (DRC) to the west. Whilst the landscape is generally quite flat, most of the country is over 1000 m (3280 ft) in altitude. Mountainous regions include the Rwenzori Mountains that run along the border with the DRC, the Virunga Mountains on the border with Rwanda and the DRC, and Kigezi in the southwest of the country. An extinct volcano, Mount Elgon, straddles the border with Kenya. The capital city, Kampala, lies on the shores of Lake Victoria, the largest lake in Africa and second-largest freshwater inland body of water in the world. Jinja, located on the lake, is considered to be the start point of the River Nile, which traverses much of the country. The varied scenery includes tropical forest, a semi-desert area in the northeast, the arid plains of the Karamoja, the lush, heavily populated Buganda, the rolling savannah of Acholi, Bunyoro, Tororo and Ankole, tea plantations and the fertile cotton area of Teso.

Gulu District is a district in Northern Uganda. The district is named after its chief municipal, administrative and commercial center, the town of Gulu. The District is bordered by Lamwo District to the north, Pader District to the east, Oyam District to the south, Nwoya District to the southwest and Amuru District to the west. The district headquarters at Gulu are located approximately 340 km (210 mi), by road, north of Uganda's capital city, Kampala. The coordinates of the district are: 02 45 N, 32 00E.

Malaria transmission occurs year-round with two peaks from May to June and from November to December following distinct rainy seasons in Northern Uganda. In addition to climate and altitude, other factors that influence malaria in the country include high human concentration near vector habitats (e.g., villages and boarding schools in proximity to marshlands or rice fields), population movement (especially from areas of low to high transmission), irrigation schemes (especially in the eastern and southern parts of the country), and cross-border movement of people (especially in the eastern and southeastern parts of the country).

3.2 UAV Tactics

Drone surveys were carried out using a DJI Phantom 4 Pro quadcopter fitted with a DJI 4 K camera for seasonal, capture point, Anopheline, larval habitat, imagery collection. The camera was composed of single-band cameras [Green, Red, Red Edge and Near Infrared—(NIR)] of 1.2 MP for multispectral imagery collection.

Multispectral mapping was conducted over 7 randomly sampled water bodies. In each water body, the drone was flown to an altitude of approximately 6ft to 25 ft, which assured a GSD of 0.02 m/pixel. A grid of 270×270 m was drawn in Pix4D and the RGB multispectral camera was set up to take an image each second during the 20-minutes flight time of the drone.

The image classification was conducted in Google Earth Engine (GEE). GEE is a cloud-based platform for planetary- scale geospatial analysis. [https://earthengine.google.com]. All classification analyses were conducted in the online Integrated Development Environment (IDE) at https://code.earthengine.google.com (repositories for data and code available in Supplementary information). All 8-band multispectral orthomosaics were uploaded to GEE assets and a supervised classification was performed using AI algorithms on the habitat capture point data.

Machine Learning (ML) in Earth Engine is supported with: (1) EE API methods in the ee.Classifier, ee.Clusterer, or ee.Reducer packages for training and inference within Earth Engine. (2) Export and import functions for TFRecord files to facilitate TensorFlow model development. Inference using data in Earth Engine and a trained model hosted on Google's AI Platform is supported with the ee.Model package.

3.3 Technical Details Data Preparation

We collected around 32 min video of the whole village using DJI4 Drone (GPS enabled) during three seasonal collection frames (dry, pre-rain and rain) during the entire year of 2019. Essentially four data gathering experiments were conducted, and each experiment was approximately 8 min long. Relevant details on our dataset are shown in Table 1. This data was used to train and validate our AI algorithm. We state

Table 1 Relevant details on our dataset of breeding habitats

Total duration of all videos	32 min
Total number of frames	1889
Total number of frames with breeding habitats	1100
Total number of frames without breeding habitats	789

clearly here that the height of the drone was chosen between 6ft to 25 ft during these experiments.

After the process of data collection, and in order to process the videos, we first extracted each frame within each video as one image. For the entire duration of 32 min the number of frames (i.e., images) extracted was 1889. Out of these, the images which contained potential anopheline larval habitats (i.e., sources of standing water) were 1100 that were subsequently annotated (localized and labeled) using *labelImg* tool (see [13]). The others did not contain any source of stagnant water. Annotated images were further verified by an expert researcher for ground truth validity. In our efforts in designing an AI algorithm for classification, we employed 70% of annotated images towards training a model, and the rest of images were used for validation. The total number of training and validation images generated after the aforementioned split were 770, and 330 respectively (for the class of images containing a potential larval habitat), and the split was similar for the other classes.

We point out that the resolution of each original image extracted was 4096 × 2160 pixels, where each pixel was a representation of a RGB color space. This was a relatively large size hence we assumed that there would be slow down training time. In order to accelerate training, without compromising accuracy, we reduced the image size by a factor of 4. This resulted in an image of size 1024 × 524 (we observed no loss in model accuracy with this reduced size). After that, we normalized the RGB value of each pixel in an image by dividing it by 255. This aided in avoiding poor contrast from the image.

Further, to increase the training images, we randomly zoomed in/zoomed out each image between 0.5 and 1.5. Doing so, helped add more robustness to our model for operating on unseen data. Via this procedure, the total number of training images (original and scaled) generated was 1540. We point out that the DJI Phantom drone provided GPS extraction capability using the notion of .SRT files containing the GPS information for each frame. Naturally, these were also extracted per frame, and they were stored in a .CSV file.

3.4 Convolutional Neural Network Based Object-Detection Algorithms to Localize the Breeding Habitats

The state of the art techniques in image recognition relies on the notion of convolutional neural networks (CNNs) [2]. We provide a brief overview here. There are two components here - feature extraction part and classification.

During feature extraction, the network performed a series of convolutions and pooling to get the critical LULC features in the image that aided in the larval anopheline habitat classification. The convolution layers extracted features from the images by performing convolution operation with the use of filters on the input images to generate the feature map. Each convolution layer had 3 dimensions (width, height and depth). Typically the image contained n filters where the filter size was *(a, b, c)* and where *a* and *b* was the width and height of the filter and *c* represented the number of color channels of an image. Each filter independently convolved on the input image that was subsequently followed by pooling to generate a feature map. The pooling layer aided in reducing the size of the feature map. After the filter size was chosen, stride size needed to be chosen. Stride size is the size of the step by which filter moves across an image. The process of convolution worked by computing a dot product between the filter and the local region of the image on which the filter was mounted. Since deep convolutional neural networks contain several convolution layers, each layer employed different filter sizes. As such, different feature maps were integrated together at the end and this acted as the final output of the convolution layer.

Subsequently, after the convolution layers, we added a few dense layers for classification. The neurons (essentially a non-linear function that takes multiple inputs and renders out a single output) in the dense layers was fully connected to all the neurons of the previous layers. The last dense layers consisted of neurons equal to the number of classes. We aimed to classify, and render the probability of each class present in the image. The one with the highest probability was predicted as the correct class for the particular image when presented for classification.

While classification of potential *An. gambiae s.l., arabiensis s.s.* and *funestus s.s.* larval habitats in an image may alone suffice in most cases, we wanted to add another feature in our design to also highlight where the predicted breeding sites was within the image (i.e., localization of the capture point object within the image). When the operator viewed the habitat localized (bounded within a solid box) more details about the size of the larval habitat was able to be inferred. Furthermore, the number of habitats present in the image were accessible. In addition, we generated a confidence metric to the operator for each bounding box which was emplaced in an image. Our solution was based on the notion of Faster R-CNN.

In the Region CNN approach, we executed several steps which were challenging. First, we trained the drone image temporally dependent datasets using a pre-trained convolutional neural network model and extracted the convolutional feature map from the last layer of the trained network. This step enabled the neural network to understand the key features within the image that separates multiple classes of objects within the image. To train towards feature recognition, the right neural network must be used and adapted. For this study, we used Inception V2 [14] as the base pre-trained convolutional network for extracting the feature maps. Inception V2 is a complex deep learning architecture which uses smart factorization methods to make the convolutions efficient in terms of computational complexity. The network performs better when the input dimensions are not changed drastically [3]. This architecture helped in achieving the same for our problem.

During Post training, we employed features maps in our image dataset that were the most reflective of the capture point larval habitats. In order to localize objects of interest in the image, we used the notion of Region CNNs (R-CNNs). To do so, a few steps needed to be executed. First, we predefined anchor aspect ratios and scaled in our images. Anchors are the rectangular boxes that are used to scan objects in the image. These were emplaced during training the neural network. For the case of our drone imagery, we set the base anchor size as [256, 256] pixels, with scaling ratios and aspect ratios as [0.25, 0.5, 0.75, 1.0, 1.50, 2.0, 2.5] and [0.25, 0.5, 0.75, 1.0, 1.50, 2.0, 2.5] respectively.

The width and height of each anchor was set as

$$width\_anchor = scale[i] * sqrt(aspect\_ratio[i])*$$
$$base\_anchor[0]height\_anchor = scale[i] * base\_anchor[1]/$$
$$sqrt(aspect\_ratio[i]),$$

where i was the index of the matrices of scales and aspect ratios. In total, 300 anchors were generated per capture point LULC image for training. Then, we fed our anchors and feature map to the region proposal network (RPN). Here, the task was to train the network to identify those boxes in the image that were indicative of larval habitats. To do so, we manually ground truthed each box prior to training as a potential anopheline larval habitat and the classified background. The RPN should be trained to identify the anchors having the relevant objects of interest (in our case, a potential breeding site) with an "objectness" score and returns the ones that most likely contains objects within the image based on the score.

For training (and validation) of the UAV habitat imaged data, the RPN was connected to a convolutional layer that had 512 output nodes. These output nodes were connected to two convolutional layers where one acted as a classifier and another as a regressor. The Classifier connected to the RPN predicts the probability of the object of interest within the anchors and the regressor estimates the tight boundary surrounding the corresponding objects using anchors. We fine-tuned the classifier by varying learning rate. We used stochastic gradient descent (SGD) solver for 50,000 iterations with a base learning rate of 2e-5, then another 25,000 iterations by reducing the base learning 2e -6 and the rest 25,000 with 2e-7 for faster convergence (Fig. 1).

3.5 Learning Rate for Each Iteration

During training, one important criteria for correctness is the loss function. Briefly, the loss function measures the learning ability of a neural network architecture. Typically, it ranges from 0 to 1 where 0 means perfect learning and 1 means no learning. The goal of training the anopheline habitat data was to minimize the loss of training data. We noticed that the default binary cross entropy loss function (which is standard) had increased after 15,000 iterations. To minimize the loss, we applied a novel focal loss function which penalized instances of false negatives which in our case was an actual larval habitat classified as background. In this research, the number of anchors containing breeding habitats was minimal in comparison to anchors that contained the background class. Hence, the classifier was biased towards background class which initially resulted in biased learning. Focal loss is an improvement over the more standard cross entropy loss, since it operates by lowering the loss of well classified cases and emphasizing the misclassified ones. We considered the following equation

Here y denoted the ground-truth of the LULC class (+1 for larval habitat and -1 for background), and, which was the UAV real time model estimated probability for the class label We quantitated focal loss. We added an additional modulating factor $(1 - Q)^{\lambda}$ to the above loss, to provide tunability, where $\lambda >= 0$. The Focal loss (FL) was defined as $FL(Q) = -(1-Q)^{\lambda}\log(Q)$.

In this research, when a potential *Anopheles* larval habitat was misclassified as background and Q was small, the modulating factor was close to 1 and this did not affect loss. When Q tends to 1, the modulating factor is close to 0 and loss for well classified examples is down-weighted. $\lambda = 2$ worked best in our case. After applying focal loss, we were able to see, as in the Fig. 2, that our goal to minimize loss was achieved. Essentially, when the loss in training data is similar to the loss in validation data (recall that validation data is not used to train the model), and when they do not decrease any further, the process of training is complete. We state here that the process of training and validation of the anopheline capture point habitat model was an iterative process. Due to space limitations, we do not (and as is typical) discuss all parameters used in each round of training and validation. The final set of parameters of the neural net archive that met our (loss) criteria were the ones presented above.

Finally, we needed to define a user-understandable metric that would reveal the quality of our neural network trained and validated for this problem. Our metric was Intersection Over Union (IoU). The IoU metric measures the correctness of a given bounding box. It is formally, the area of the intersection of the predicted box and ground truth box divided by the area of union of the predicted box and the ground

Fig. 1 The learning rate for each iteration

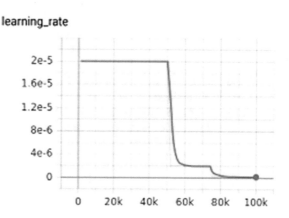

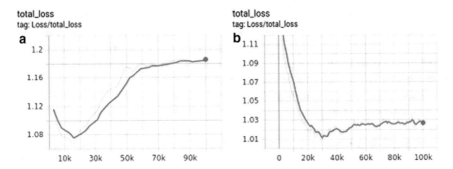

Fig. 2 **a** and **b**: Total loss graph without and with focal loss

truth box. In our design, the IoU threshold was set as at-least 0.7. When the IoU is 1, then that a perfect classification and emplacement of the bounding box occurred. Lower IoU values indicate poorer performance.

By computing the True Positives, False Positives and False Negatives for each habitat classification in the validation set, we were able to integrate a final metric called mean Average Precision or *mAP*. Denoting AP as the Average Precision (AP) in finding the area under the precision-recall curve of each class, the mean Average Precision or *mAP* score was calculated by taking the mean AP over all LULC classes and/or over all IoU thresholds. Note that Precision and Recall are standard metrics in binary classification problems. The Precision was defined as the ratio of True Positives to the Sum of True Positives and False Positives. The Recall was subsequently defined as the ratio of True Positives to the Sum of True Positives and False Negatives. A high Precision and Recall indicate a more accurate classifier.

The *mAP* of all images in our validation set was determined to 0.87 Note that if we had set the IoU with lower thresholds (i.e., less than 0.7), the *mAP* was higher.

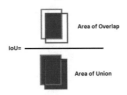

We emphasize here that we trained our Faster R-CNN model using the annotated images to detect and locate sources of breeding habitats. (Figure 3 a, b). It took close 10 h for training and validating the model using a graphic processing unit (GPU) cluster. The cluster has 4 nodes of GeForce GTX TITAN X each with 12 GB memory.

Fig. 3 An instance of a predicted anopheles larval habitat akonyibedo by our AI algorithm

4 Results

In order to test the correctness (or accuracy) of our neural network habitat model, we tested with completely unseen images taken from drones. By unseen, we mean that the images that were fed into the neural network for classification and localization of larval habitats which were never used during training or validation (details of which were elaborated above). Our algorithm performed optimally here, and we were confident that our model was one of high fidelity.

For testing, we once again flew our drone over new areas (different from ones used to finalize the neural network). This time, we took seven videos with total duration of 15 min. The total number of frames extracted was 959 with 60% of them containing at-least one potential larval habitat (some of them repetitive in that the same breeding spot was present in multiple frames).

Each frame was fed into our model for classification and localization, and we derived excellent results. From Figs. 3, 4 and 5 (just representative instances of our testing image dataset), we can see that breeding habitats were identified accurately. The probabilities of detection by our neural network (i.e., the confidence it had in making the prediction) was close to 0.99 in most images. Not a single False Negative

Fig. 4 An instance of a predicted *Anopheles* larval habitat akonyibedo by our AI algorithm

Fig. 5 An instance of a predicted *Anopheles* larval habitat akonyibedo by our AI algorithm

existed in our testing dataset, meaning that every potential source of stagnant water (i.e., larval habitat) was correctly classified and localized, and there was no false positives.

For practical deployment, once a frame was predicted to have at-least one anopheline larval habitat, we extracted its GPS location from the associated.csv file (which was provided by the drone). The final output to the operator was provided as an image with bounding boxes and its GPS location was able to locate the habitats in the form of a simple smartphone app.

These habitat sites were able to be subsequently be classified as productive or not for prioritizing seasonal, breeding sites, for implementing larval control strategies by overlaying an Anopheline, RGB analog, video larval habitat signature over the UAV, real time, geosampled, LULC images.

The sensitivity and specificity of the drone signals at identifying positive and dry habitats were subsequently evaluated in blinded experimental studies in a natural setting followed by extensive ground-truthing of the UAV sampled, real time, LULC classified, model outputs employing GCPs which had a positional accuracy 0.178 m. All or 100% of the capture points from 90 (unique) larval habitats forecasted by the model in Akonyibedo village were identified, and were found to contain *Anopheles* larvae.

Our results reveal that at 100% active breeding sites at baseline, on average, over 18 females *Anopheles* mosquitoes were found per house Pyrethrum Spray Collection (PSC) sprayed. On destruction of 85% breeding sites employing Seek and Destroy, a monitoring PSC revealed a vector abundance to 1 female *Anopheles* mosquitoes per house in 30 days. This means that if all breeding sites where destroyed, there would be an effective vector reduction to zero indoor resting female anopheles mosquitoes per household

5 Discussion/Conclusion

The AI component of our technique was the successful training of a Faster R-CNN (Region-based Convolutional Neural Network) model using all annotated images from the drone video to detect and locate sources of anopheline aquatic breeding locations at the intervention, epi-entomological site. The R-CNN model essentially performed two tasks, classification and regression. The classification component found the object of interest (in our case, a potential larval habitat) within an image, and the regression components emplaced a tight bounding box. Our final metric to assess accuracy was *mAP*, which is a function of another metric called IoU that essentially compared the overlap amongst the ground truth and predicted bounding boxes after classification and localization. For a relatively high IoU threshold of 0.7, the *mAP* value during validation was 0.87. With more training data, the accuracy will only improve further for targeting seasonal, anopheline larval habitats in agro-village pastureland ecosystems throughout peri- domestic pastureland environments in Uganda.

The application of machine learning algorithms to support public health intelligence and shape intervention strategies has been discussed. Sensemaking lies at the heart of location intelligence. The sensemaking applies not only to the targeted interventions to eliminate vectors, but also strategic sensemaking that contextualizes this intervention as part of a more holistic/systemic and strategic intervention encompassing a myriad of coordinated interventions across the disaster management spectrum (mitigation, preparedness, response, recovery).

References

1. Jacob B, Mwangangi JM Mbogo CB, Novak RJ (2011) A taxonomy of unmixing algorithms using decomposing a quickbird visible and near infra-red pixel of an anopheles arabiensis habitat. Open Remote Sens 17(3):11–24
2. Krizhevsky A, Sutskever I, Hinton GE (2012) Imagenet classification with deep convolutional neural networks. In: Advances in neural information processing systems, pp 1097–1105

3. Lin TY, Goyal P, Girshick R, He K, Dollár P (2017) Focal loss for dense object detection. In: Proceedings of the IEEE international conference on computer vision 2017, pp. 2980–2988
4. WHO (2020) https://www.who.int/news-room/fact-sheets/detail/malaria
5. Weick KE (1995) Sensemaking in organizations. Sage, Thousand Oaks, CA
6. Ancona DL (2011) SENSEMAKING framing and acting in the unknown in handbook of teaching leadership. Sage Publications, Inc, Thousand Oaks, CA, pp 3–20
7. Crayne MP, Medeiros KE (2020) Making sense of crisis: Charismatic, ideological, and pragmatic leadership in response to COVID-19. American Psychologist
8. Weick KE, Sutcliffe KM, Obstfeld D (2005) Organizing and the process of sensemaking. Organ Sci 16:409–421
9. Jacob B, Novak RJ (2015) Integrating a trimble recon X 400 MHz Intel PXA255 Xscale CPU® mobile field data collection system using differentially corrected global positioning system technology and a real-time bidirectional actionable platform within an ArcGIS cyberenvironment for implementing malaria mosquito control. Adv Remote Sens 3(3):141–196
10. Jacob BG, Novak RJ (2019) Efficaciously targeting unknown *Anopheles minimus* orthomosi-acked eco- georeferenceable capture points employing photogrammetric Rayleigh optical depths as a function of sub-meter resolution semi-infinite anisotropical azimuthally krigable asymmetrical scattering signature spectroscopic frequencies under disproportionate solar exoatmospheric irradiance conditions in eigenspace in a semi- autonomous unmanned aircraft ArcGIS real time dashboard Principal Component Analysis. Ann Biostatistics Biometric Appl (In press 2019)
11. Maloof M. Artificial intelligence: an introduction, p. 37" (PDF). *georgetown.edu*
12. Ren S, He K, Girshick R, Sun J. Faster r-cnn (2015) Towards real-time object detection with region proposal networks. In: Advances in neural information processing systems, pp 91–99
13. GitHub-Tzutalin L (2019) LabelImg is a graphical image annotation tool and label object bounding boxes in image. Accessed 21 Dec 2019. https://github.com/tzutalin/labelImg
14. Szegedy C, Vanhoucke V, Ioffe S, Shlens J, Wojna Z (2016) Rethinking the inception archi-tecture for computer vision. In: Proceedings of the IEEE conference on computer vision and pattern recognition 2016, pp 2818–2826

Exploring the Opioid Crisis Through Systems Thinking and Participative Model Building: An Experiential Learning Event

Anthony J. Masys, Colleen C. Reiner, Nancy Ramirez, and Haru Okuda

Abstract The impact of the opioid crisis is significant resulting in not only rising mortality and morbidity but also significant social and economic costs. As detailed in Felbab-Brown et al. (The Opioid Crisis in America Domestic and International Dimensions. Brookings Paper Series, [1]), 'Since 2000, there have been 400,000 opioid-involved deaths in the U.S. In the United States of America the number of people dying from opioid overdose increased by 120% between 2010 and 2018'. Stemming from a 2018 survey, it is estimated that upwards of 2.35 million Americans are suffering from an Opioid Use Disorder (OUD). Today's public health crisis associated with opioid use is characterized by complexity requiring more than one discipline to address the issues. Healthcare professionals must therefore work in inter-professional teams in order to better communicate and address these complex and challenging needs (Bridges et al. in Medical Education 16, [2]). Systems thinking is an approach to better understand the interconnectivity and interdependencies associated with public health issues including how to intervene to improve individual and population health. Inter-professional experiences thereby support development of skills facilitating opportunities to become '…collaborative inter-professional team members who show respect and positive attitudes towards each other and work towards improving patient outcomes' (Bridges et al. in Medical Education 16, [2]). This chapter examines the application of systems thinking and participative model building in an Interprofessional Education (IPE) event held in November 2019 by USF Health, University of South Florida. The IPE event focused on the opioid crisis and was comprised of a learning journey of concept development, collaborative model building and shared sensemaking across healthcare professionals.

Keywords Interprofessional education (IPE) · Opioid crisis · Systems thinking · Participative model building

A. J. Masys (✉)
College of Public Health, University of South Florida, Tampa, FL, USA

C. C. Reiner · N. Ramirez · H. Okuda
Center for Advanced Medical Learning and Simulation (CAMLS), University of South Florida, Tampa, FL, USA

© The Author(s), under exclusive license to Springer Nature Switzerland AG 2021   149
A. J. Masys (ed.), *Sensemaking for Security*, Advanced Sciences and Technologies for Security Applications, https://doi.org/10.1007/978-3-030-71998-2_9

1   Introduction

The impact of the opioid crisis is significant resulting in not only rising mortality and morbidity but also significant social and economic costs. As detailed in Felbab-Brown et al. [1], 'Since 2000, there have been 400,000 opioid-involved deaths in the U.S. In the United States of America the number of people dying from opioid overdose increased by 120% between 2010 and 2018'. Stemming from a 2018 survey, it is estimated that upwards of 2.35 million Americans are suffering from an Opioid Use Disorder (OUD). For context, more people in the United States died from overdoses involving opioids in 2017 than from HIV- or AIDS-related illnesses at the peak of the AIDS epidemic [3].

Today's opioid public health crisis is characterized by complexity requiring more than one discipline to address the issues. Systems thinking is an approach to better understand the interconnectivity and interdependencies associated with public health issues including how to intervene to improve individual and population health. Many of the challenges within public health and clinical domains can be described as complex, requiring an interdisciplinary and multidisciplinary lens.

2   Systems Thinking and Sensemaking

Weick [4] defines sensemaking as simply "the making of sense". As described in Ancona [5] '…it includes the process of "structuring the unknown" … 'to comprehend, understand, explain, attribute, extrapolate, and predict'. Further [5] argues that 'sensemaking is the activity that enables us to turn the ongoing complexity of the world into a "situation that is comprehended explicitly in words and that serves as a springboard into action" [6]'. The sensemaking process individuals undertake as they try to understand the world around them is rooted in our mental models (our beliefs and assumptions about how the world works) and is leveraged/utilized in an attempt to interpret the significance of shared events and experiences. Thus, sensemaking involves coming up with plausible understandings and meanings; testing them, challenging mental models and constructs and refining understandings.

In short, sensemaking refers to the processes of interpretation and meaning creation that we use to reflect on and interpret events to produce intersubjective accounts [4]. In one context and within the scenario of the IPE event described, sensemaking can be likened to making a map of a situation (mental model) characterized by unknowns and complexity. The multiple perspectives garnered from a systems mapping approach in a team environment creates an emerging picture that becomes more comprehensive through data collection, action, experience, and conversation.

As described in Masys [7], systems thinking is both a worldview and a process in the context that it informs one's understanding regarding a system and can be used as an approach in problem solving [8]. As a cross-disciplinary domain, systems thinking spans from the physical sciences and engineering to the social sciences, health sciences, humanities and fine arts. Because of this feature of systems thinking, there is no universally agreed definition of a 'system' that satisfies all domains, although they may share similar defining characteristics [9].

As described in Masys [7], Anderson and Johnson [10] define a system as:

> ...a group of interacting, interrelated, or interdependent components that form a complex and unified whole. A system's components can be physical objects ...and can also be intangible, such as processes; relationships, policies, information flows; interpersonal interactions; and internal states of mind such as feelings, values, and beliefs.

Systems thinking, according to [11] 'is a discipline for seeing wholes. It is a framework for seeing interrelationships rather than things, for seeing patterns of change rather than static snapshots'. As a worldview, systems thinking recognizes that systems cannot be addressed through a reductionist approach that reduces the systems to their components.

Systems thinking purports that, although events and objects may appear distinct and separate in space and time, they are all interconnected. Senge [11] remarks that, because the world exhibits qualities of wholeness, our investigation of it should stem from a paradigm of the whole.

For healthcare professionals, systems thinking seeks to increase their ability to understand the actors, interconnectivity and interdependencies and stress test our mental models by asking "what-if" questions about possible future behaviours of systems that contribute to individual and population health. Hence, systems thinking provides insights to transform healthcare systems in such ways that they will behave as desired, generating wanted outcomes and creating intended value [12]. This is critical in the design of public health strategies recognizing the potential for unintended consequences.

Understanding the interdisciplinary and multidisciplinary nature of the opioid crisis, sensemaking through systems thinking becomes a powerful approach to bring different perspectives to the problem space and includes such applications as system mapping and participative model building. In so doing, systems thinking provides a lens to better understand and continuously test and revise our understanding of public health issues.

3 Interprofessional Education (IPE)

Today's public health and clinical problem spaces are characterized by complex health needs requiring an interdisciplinary approach '... in order to better communicate and address these complex and challenging needs' [2]. Interprofessional experiences thereby support development of skills facilitating opportunities to become '...collaborative interprofessional team members who show respect and positive attitudes towards each other and work towards improving patient outcomes [2]. Interprofessional collaborative practice has been defined as a process which includes communication and decision-making, enabling a synergistic influence of grouped knowledge and skills [2]. Such interactions reflect a blending of professional cultures to improve the quality of health care whether at the clinical level or public health community, regional, national or global level. Such participatory collaborative and coordinated approach is critical in developing shared decision making and situation awareness around clinical and public health issues.

Sensemaking in this context is a key element pertaining to diagnosis, prognostication, and treatment whether at the clinical level or public health community level and is often distributed across professional boundaries. As such, collaboration and solution navigation skills training within the context of an inter-professional experience (IPE) improves not only sensemaking skills but also teamwork, listening skills, and reflective and analytical thinking [13].

Interprofessional education is an approach to develop healthcare students for future interprofessional teams. Students trained using an IPE approach are more likely to become collaborative interprofessional team members who show respect and positive attitudes towards each other and work towards improving patient outcomes [2].

IPE has been defined as 'members or students of two or more professions associated with health or social care, engaged in learning with, from and about each other'. IPE provides an ability to share skills and knowledge between professions and allows for a better understanding, shared values, and respect for the roles of other healthcare professionals. The desired end result is to develop an interprofessional, team-based, collaborative approach that improves patient and community outcomes and the quality of care [2]. As a collaborative practice, IPE supports enhanced communication and decision-making, enabling a sharing of group knowledge and skills.

In this chapter, we describe an IPE experience that focuses on opening the 'black box' of the opioid crisis. This was achieved by applying systems thinking and participative model building to access and share tacit knowledge and to surface assumptions, beliefs and generate a collaborative view of the opioid problem and solution space.

4   Discussion

4.1   Understanding the Opioid Crisis

The Opioid issue can be considered a national security crisis as well as a public health crisis. The morbidity and mortality from opioid misuse, abuse, and overdose continues to rise in the United States, creating a crisis for patients, families, and communities. As detailed in Felbab-Brown et al. [1], 'Since 2000, there have been 400,000 opioid-involved deaths in the U.S. In the United States of America the number of people dying from opioid overdose increased by 120% between 2010 and 2018.

As described in CDC [14], this rise in opioid overdose deaths can be outlined in three distinct waves.

1. The first wave began with increased prescribing of opioids in the 1990s, with overdose deaths involving prescription opioids (natural and semi-synthetic opioids and methadone) increasing since at least 1999.
2. The second wave began in 2010, with rapid increases in overdose deaths involving heroin.
3. The third wave began in 2013, with significant increases in overdose deaths involving synthetic opioids, particularly those involving illicitly manufactured fentanyl. The market for illicitly manufactured fentanyl continues to change, and it can be found in combination with heroin, counterfeit pills, and cocaine.

The breadth, depth, and dynamic evolution of the opioid crisis have researchers and healthcare professionals exploring a plethora of strategies to reduce the supply, inappropriate use, and harm caused by prescription opioid analgesics and illicit opioids. Bringing multiple voices and perspectives to the problem space sheds light on the underlying drivers and unintended consequences of current interventions that are shaping the trajectory of the crisis.

Dasgupta et al. [15] describe in detail the challenges associated with a root cause analysis associated with the opioid crisis. Their study reveals an inherent complexity associated with the problem space that spans individual, societal, environmental, organizational and regulatory issues. Hence interdisciplinary dialogue facilitates a move beyond just a 'siloed' perspective of the crisis and reveals the structural and social determinants of health framework thereby bringing a more holistic understanding to the public health challenges.

4.2 Experiential Learning: Case Study

As described by [16] '...experiential learning means learning from experience or learning by doing. Experiential education first immerses learners in an experience and then encourages reflection about the experience to develop new skills, new attitudes, or new ways of thinking' (cited in Schwartz [17]. As cited in Culhane et al. [18] 'Kolb [19] suggests that learning happens when meaning making of experiences occurs. The experiential cycle depicts meaning making consisting of having concrete experiences, reflecting on those experiences, conceptualizing, and experimenting [19]. It entails creating an opportunity and environment whereby knowledge is created through the 'transformation of experience' [19].

The University of South Florida (USF) enables this "transformation of experience" through the design of its interprofessional USF Health, a unique organizational structure within the university that blends its healthcare-focused academic disciplines with their practicing physicians group. This interconnected network encourages experiential learning opportunities for students that integrate patient care, education, and research into their curriculum.

Interprofessional education is a core value within USF Health noting that regardless of discipline or unique expertise, healthcare professionals benefit by learning together and working alongside colleagues from other professions. This interprofessional collaboration is considered to be an essential component in tackling some of the most pervasive health challenges in our society whether individual patient, community health or more widely- public health.

USF Health hosted its Annual Interprofessional Education Day (IPE Day) to promote collaboration across health disciplines. Each year this cross-campus initiative includes students from 8 different academic disciplines, including: athletic training, nursing, medicine, public health, pharmacy, physician assistant, physical therapy and rehabilitation, and behavioral and community sciences. To effectively facilitate an event of this scale, leadership across USF Health designated a date on their academic calendar for the singular purpose of promoting interprofessional engagement across schools and programs. By including such a diversity of health professions, IPE Day maximizes the potential to learn from other professions with other professions.

Each year, USF Health's IPE Day examines a different theme or topic with a dynamic lecture series open to the public, followed by an experiential learning activity for USF students. The day-long event in 2019 focused on the opioid crisis, and was divided into morning and afternoon sessions. Over 800 guests participated in the morning session which consisted of keynote addresses by policy makers, educators, and medical experts, along with panel discussions. The learning objectives included (a) examining the impact of IPE collaboration on the opioid crisis among individuals, families and communities, (b) identifying the interconnected relationships between healthcare systems, and their connection to the opioid crisis in our community, and (c) recognizing the importance of IPE interventions to improve patient-centered care at different levels within the system.

The 2019 afternoon session was by invitation only and contained interactive activities for the 120 students and faculty that were selected to represent the 8 disciplines from across USF Health. The 2019 session introduced a systems-level, critical thinking activity in which participants were encouraged to consider the complex, interconnected relationships between various societal factors and opioid addiction. This interprofessional approach to the conversation provided over 100 future healthcare providers with unique skills to help them address urgent healthcare crises from a systems-level framework.

To ensure the academic rigor of the activity, USF Health implemented a validated assessment technique developed by King et al. (2016) which included pre & post confidence testing, combined with a specialized rubric developed by USF Health to evaluate our dynamic, systems-mapping activity. Our team saw a large increase in participant scores relating to IPE competencies as measured by the standardized mean difference following the completion of the afternoon activity. By examining the impact of substance use disorder (SUD) among individuals, families and communities during an IPE case scenario, participants were able to formulate research, policy, and/or clinical practice recommendations designed to mitigate the impact of SUD by leveraging interprofessional collaborative strategies.

4.3 Opening the Black Box of the Opioid Crisis: IPE, Complexity and Systems Thinking

Many of the challenges within public health and clinical domains such as that associated with the opioid crisis can be described as complex. A complex system has the following characteristics as discussed in Snowden and Boone [20]:

- It involves large numbers of interacting elements.
- The interactions are nonlinear, and minor changes can produce disproportionately major consequences.
- The system is dynamic, the whole is greater than the sum of its parts, and solutions can't be imposed; rather, they arise from the circumstances. This is frequently referred to as emergence
- The system has a history, and the past is integrated with the present; the elements evolve with one another and with the environment; and evolution is irreversible.
- Though a complex system may, in retrospect, appear to be ordered and predictable, hindsight does not lead to foresight because the external conditions and systems constantly change.
- Unlike in ordered systems (where the system constrains the agents), or chaotic systems (where there are no constraints), in a complex system the agents and the system constrain one another, especially over time. This means that we cannot forecast or predict what will happen.

Following the morning IPE session which included guest speakers and panel discussions regarding the Opioid crisis in the US, a systems thinking primer was held to introduce the methods of systems mapping and participative model building. A key message from the primer was that systems thinking involves thinking in layers, interconnectivity and and relationships. The systems thinking paradigm challenges the traditional linear thinking approach with its tendency to compartmentalize solution options and minimize recognition of relationships between solutions and their elements resulting in unintended consequences.

Systems thinking requires a shift in our perception of the world around us. In order to build a new multidimensional thinking framework, we need to discover the dynamics and interconnectedness of the systems at play. This is where systems mapping tools come in—they provide an exploration of the system, communicate understanding, and allow for the identification of knowledge gaps, intervention points, and insights.

Visualization, visual thinking, descriptive and generative diagramming and participative model building are applied in processes of sensemaking as well as for communication and involves participation and collective production of information [21].

In developing a systems mapping, all the elements, agents, actors, nodes, components are mined, the key part is drawing the connections and relationships between them. This is where you can start to tease out all the non-obvious parts and seek to

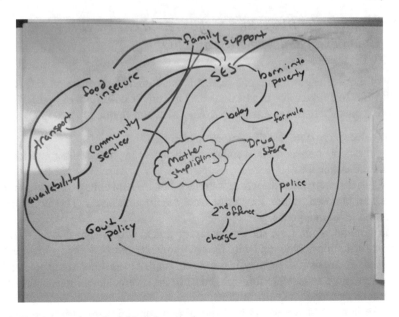

Fig. 1 System mapping of opioid crisis scenario

develop a more holistic view of the system you are exploring. From this, insights can be drawn, and potential interventions (leverage) points can be identified.

One of the key elements of systems mapping and systems thinking is to suspend the need to solve and embrace the complexity of the system. Relationships are messy, so the systems map should also reflect this. Suspending judgment and the need to have a 'right' or 'wrong' answer to the exploration is key. As an IPE exercise, the participants worked in teams developing a systems map of the opioid crisis facilitating the emergence of connections and relationships across the health disciplines. Figure 1 illustrates a systems mapping of a scenario associated with the workshop.

The systems mapping begins with a central problem or theme and through interdisciplinary dialogue, branches are extended from the central theme highlighting different perspectives of the problem space. This presents a more holistic view of the opioid crisis by embracing participative interdisciplinary dialogue.

The systems mapping is used to facilitate dialogue as well as:

- To make sense of complexity
- Better see and understand system
- Clarify relationships between system components
- Challenge assumptions and mental models
- Identify root causes
- Understand feedback loops and patterns
- Identify knowledge gaps
- To engage various health disciplines and leverage different perspectives
- Communicate complexity

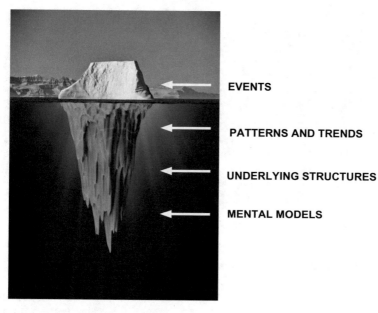

EVENTS

PATTERNS AND TRENDS

UNDERLYING STRUCTURES

MENTAL MODELS

Fig. 2 The Iceberg model of systems thinking

• Co-create shared understanding of the system

System mapping provides a methodology that explores the various dimensions of the iceberg model (Fig. 2). When we explore the opioid crisis through the systems lens, we move beyond an event description of the crisis to uncovering the trends, underlying structures and mental models that shape our understanding of the crisis and shape our design of intervention strategies. The systems thinking approach applied in the IPE allowed the participants to explore the opioid crisis from different perspectives revealing the dimensions illustrated in the Iceberg model.

Systems mapping as a participatory mixed methodology, enables diverse participant groups to develop shared conceptual frameworks. The results are generative, encouraging shared meaning and interprofessional learning while preserving individuality and diversity.

Following the systems mapping exercise, the interprofessional teams then created a 3D model: a representation of the problem and solution space associated with the opioid crisis. The participative model building exercise resulted in a rich metaphorical construct of their diverse understandings of their complex situation (Fig. 3).

Participative model building is an approach that is widely used to build the capacity of practitioners to think in a systems way [22]. As noted in Siokou et al. [22], it facilitates '…a means of engaging diverse stakeholders in a process for jointly understanding and addressing complex issues, such as disease prevention and health promotion'.

Fig. 3 Participative model building results

In the participative model building process, participants, leveraging the systems mapping results, use modeling to further describe the problem, to identify, develop and test solutions, and to inform the decision making and actions of the group. In this way the participative model building was a purposeful learning process for action that engages the implicit and explicit knowledge of stakeholders to create formalized and shared representations of reality [23].

The systems mapping and participative model building '…is about acute perception, or better, about re-perception—becoming free of old perceptions and prejudices at the same time' [24].

Through facilitated discussions, participants were encouraged to construct their own meanings and explore diverse interpretations in the development of the system map and model. Systems thinking provides an opportunity for participants to develop multiple ways to read and understand situations. As an experiential learning model, system mapping and model building promotes an innovative way to enhance the experience of dialogue and sensemaking. It also fosters a learning environment for participants to explore multiplicity, imagination, and uncertainty.

Sensemaking in this IPE emerged from sharing diverse sets of knowledge from different people that cut across various skills and expertise domains creates a greater variety of perspectives and ideas, thereby increasing the opportunities for individual learning. From a design thinking perspective [25], team members bring in different pieces of information and contribute to the generation of a rich repository of ideas for problem solving. Knowledge combining refers to the bridging of existing ideas, proposed solutions, and posts in novel and creative ways by crowd members to generate new ideas. When aspects of knowledge combining are enabled, such as the ability to edit and interact, then the collective intelligence of the team is unlocked. Individual team members can work through the difficulties and sticking points associated with complex problems and draw on the diversity of expertise (Fig. 4).

Fig. 4 Video illustrating the systems mapping activity implemented as part of USF Health's IPE Day 2019. https://www.youtube.com/embed/q-fyFTLrx1c?feature=oembed

5 Conclusion

Interprofessional education is the key if we want to make a difference in this community.

Events like IPE Day bring professionals and experts from across disciplines together, so that we can tackle the opioid crisis in a meaningful way, from a 360 degree approach. We must all recognize that opioid addiction is a national crisis and it does not discriminate between race, sex, or socioeconomic class. Drug addiction destroys families and lives. In the final analysis, effectively addressing the opioid crisis is going to take our entire community. Tampa Mayor, Jane Castor[1]

Complexity is the core feature of many public health problems today. We are dealing with a world characterized by nonlinearities, tipping points, and asymmetrical relations where a small cause can have a big effect. In a systems approach, local issues need more global solutions.

As described in Hynes et al. [26], 'applying a systemic lens to complex problems can help map the dynamics of the system, explore the ways in which the relationships between system components affect its functioning, and ascertain which interventions can lead to better results. Thus, systems thinking can help facilitate innovation in the health sector and systems thinking tools and methods could be the solution for twenty-first century missions, where the public problems and purposes are shifting and methods to adapt the institutions need to also reflect the shifts'.

---

[1]Worth, S. (2021). *It will take a cross—discipline approach to halt the opioid crisis, experts at IPE Day say - USF Health News*. USF Health News. Retrieved 11 Jan 2021. from https://hscweb3.hsc.usf.edu/blog/2019/11/15/it-will-take-a-cross-discipline-approach-to-halt-the-opioid-crisis-experts-at-ipe-day-say/.

Systems thinking applied to a clinical and public health crisis has been used as a 'sensemaking' tool to make interconnectedness visible. The application of systems mapping and participative model focuses attention on challenging our assumptions we bring to the table and through the process of interdisciplinary discovery, participative and immersive learning and invention reveal novel insights. The result was the emergence of a rich metaphorical understanding of the complex situation associated with the opioid crisis. The process of systems mapping and participative model building created an inclusive approach to understanding the problem space that harnessed creativity and multivocality.

As an experiential learning model, system mapping and participative model building promotes an innovative way to enhance the experience of dialogue and sense-making. It also fosters a learning environment for participants to explore multiplicity, imagination, and uncertainty and provides innovative experiences for students to prepare them to work in complex healthcare settings.

References

1. Felbab-Brown V, Caulkins JP, Graham C, Humphreys K, Pacula RL, Pardo B, Reuter P, Stein BD, Wise PH (2020) The opioid crisis in America domestic and international dimensions. Brookings paper series| June 2020
2. Bridges DR, Davidson RA, Odegard PS, Maki IV, Tomkowiak J (2011) Interprofessional collaboration: three best practice models of interprofessional education. Med Educ 16. 10.3402/meo.v16i0.6035. https://doi.org/10.3402/meo.v16i0.6035
3. Deweerd S (2019) The natural history of an epidemic Understanding how the opioid epidemic arose in the United States could help to predict how it might spread to other countries. S10| Nature 573, 12 Sept 2019
4. Weick KE (1995) Sensemaking in organizations. Sage, Thousand Oaks, CA
5. Ancona DL (2011) SENSEMAKING framing and acting in the unknown. In: Handbook of teaching leadership. Sage Publications, Inc, Thousand Oaks, CA, pp 3–20. https://www.sagepub.com/sites/default/files/upm-binaries/42924_1.pdf
6. Weick KE, Sutcliffe KM, Obstfeld D (2005) Organizing and the process of sensemaking and organizing. Organization Science 16(4):409–421
7. Masys AJ (2010) Fratricide in air operations: opening the black box- revealing the social. Ph.D. Dissertation, June 2010, University of Leicester, UK
8. Edson R (2008) Systems thinking. applied: a primer. ASysT Institute, http://www.asysti.org/OurModules/SharedDocs/HistoryDocView.aspx?ID=373
9. Checkland P (1981) Systems thinking, systems practice. Wiley, Chichester
10. Anderson V, Johnson L (1997) Systems thinking basics: from concepts to causal loops. Pegasus Communications Inc, Waltham, MA
11. Senge P (1990) The fifth discipline: the art and practice of the learning organization. Doubleday Currency, New York
12. Blokland P, Reniers G (2020) Concept paper safety science, a systems thinking perspective: from events to mental models and sustainable safety. Sustainability 12:5164
13. Bramstedt KK (2016) IMAGES OF HEALING AND LEARNING The use of visual arts as a window to diagnosing medical pathologies. AMA J Ethics®. 18(8):843–854
14. CDC (2020) https://www.cdc.gov/drugoverdose/epidemic/index.html
15. Dasgupta N, Beletsky L, Ciccarone D (2018) Opioid crisis: No easy fix to its social and economic determinants. AJPH 108(2). https://www.ncbi.nlm.nih.gov/pmc/articles/PMC5846593/pdf/AJPH.2017.304187.pdf

16. Lewis LH, Williams CJ (1994) In: Jackson L, Caffarella RS (eds) Experiential learning: a new approach. Jossey-Bass, San Francisco, pp 5–16
17. Schwartz M (2012) Best practices in experiential learning. https://www.ryerson.ca/content/dam/experiential/PDFs/bestpractices-experiential- learning.pdf
18. Culhane J, Niewolny K, Clark S, Misyak S (2018) Exploring the intersections of interdisciplinary teaching, experiential learning, and community engagement: a case study of service learning in practice. Int J Teach Learn High Educ 30(3):412–422
19. Kolb DA (1984) Experiential learning: experience as the source of learning and development. englewood cliffs. Prentice-Hall, New Jersey
20. Snowden DJ, Boone ME (2007) A leader's framework for decision making. Harvard Business Review 85(11):68–77
21. Sevaldson B (2011) Giga-mapping: visualisation for complexity and systems thinking in design. In: Nordic design research conference 2011 Helsinki
22. Siokou C, Morgan R, Shiell A (2014) Group model building: a participatory approach to understanding and acting on systems. Public Health Res Practice. 25(1):
23. Voinov A, Jenni K, Gray S, Kolagani N, Glynn PD, Bommel P, Prell C, Zellner M, Paolisso M, Jordan R, Sterling E, Schmitt Olabisi L, Giabbanelli PJ, Sun Z, Le Page C, Elsawah S, BenDor TK, Hubacek K, Laursen BK, Jetter A, Basco Carrera L, Singer A, Young L, Brunacini J, Smajgl A (2018) Tools and methods in participatory modeling: Selecting the right tool for the job. Environmental Modelling and Software. https://doi.org/10.1016/j.envsoft.2018.08.028
24. Kahane A (2012) Transformative scenario planning. Berrwtt-Koehler Publishers Inc., San Francisco
25. Mount M, Round H, Pitsis TS (2020) Design thinking inspired crowdsourcing: toward a generative model of complex problem solving california management review 62(3):103–120
26. Hynes W, Lees M, Müller J (eds) (2020) Systemic thinking for policy making: the potential of systems analysis for addressing global policy challenges in the 21st century, New Approaches to Economic Challenges. OECD Publishing, Paris. https://doi.org/10.1787/879c4f7a- en

Vulnerability Analysis to Support Disaster Resilience

Blake Scott and Anthony J. Masys

Abstract Hurricane Maria hit the United States (U.S.) territory of Puerto Rico on September 20th, 2017 as a Category 4 storm. An estimated 2,975 people died in Puerto Rico due to the storm and in many ways the island was devastated from the disaster. The U.S. federal government mounted a large-scale disaster response to mitigate damage by addressing basic needs. However, two years after Hurricane Maria, Puerto Ricans were still reporting difficulties in daily life, including access to health care. Puerto Rico's higher poverty levels, aging population, and struggle to provide social services have made the effects of Hurricane Maria that much more profound for its residents. The purpose of this chapter is to understand the irreparable damage to Puerto Rico's infrastructure, population, and health care system caused by Hurricane Maria compared to how other parts of the US have recovered from similar overwhelming natural disasters. Additionally, this chapter will evaluate how these incidents impact security and how sensemaking can be used to increase disaster resilience.

Keywords Puerto rico · Hurricane maria · Resilience · Vulnerability · Health care system

1 Introduction

Hurricane Maria hit the United States (U.S.) territory of Puerto Rico on September 20th, 2017 as a Category 4 storm. An estimated 2,975 people died in Puerto Rico due to the storm and in many ways the island was devastated from the disaster. The U.S. federal government mounted a large-scale disaster response to mitigate damage by addressing basic needs. However, two years after Hurricane Maria, Puerto Ricans were still reporting difficulties in daily life, including access to health care.

B. Scott · A. J. Masys (✉)
College of Public Health, University of South Florida, Tampa, USA

A. J. Masys
International Centre for Policing and Security, University of South Wales, Newport, Wales, UK

© The Author(s), under exclusive license to Springer Nature Switzerland AG 2021
A. J. Masys (ed.), *Sensemaking for Security*, Advanced Sciences and Technologies for Security Applications, https://doi.org/10.1007/978-3-030-71998-2_10

Weick and Sutcliffe [68] argue that, 'Unexpected events often audit our resilience' ...everything that was left unprepared becomes a complex problem, and every weakness comes rushing to the forefront'. This points to the reflective questions:

Was the devastating Hurricane Maria an unexpected event?
Was it a 'black swan' [54]?
Was it 'the elephant in the room'?

Taleb [55] argues that 'Not seeing a tsunami or an economic event coming is excusable; building something fragile to them is not'. The devastating impact of the hurricane on Puerto Rico lends itself to an analysis of the inherent vulnerabilities in the communities and societal critical infrastructure.

Societal critical infrastructure is a complex system replete with interdependencies and interconnectivity. It permeates all avenues of societal activities: from water and sanitation, to energy and electricity, to the provision of public health and the economy. Linkov et al. [21] argues that inherent vulnerabilities in critical infrastructure and societal systems can be exposed during events that trigger cascading systemic failures. 'Certain triggers of systemic threats can be violent and forceful, jarring a relatively stable and sustainable system into an altogether different configuration' [21]. Cascading events and disaster impacts on societal infrastructure are also discussed in Masys [27–29].

Systemic threats represent growing challenges nationally, regionally and globally. The experience of COVID-19 on national and global societal systems certainly reflects the inherent fragility in our interdependent systems. As referenced in Linkov et al. [21], Centeno et al. [5] described systems threats as, '...the threat that individual failures, accidents, or disruptions present to a system through the process of contagion. The International Risk Governance Center contends that systemic threats arise when "systems [...]are highly interconnected and intertwined with one another", where a disruption to one area triggers cascading damages to other nested or dependent nodes (IRGC [8]). Further, IRGC [8] states that "external shocks to interconnected systems, or unsustainable stresses, may cause uncontrolled feedback and cascading effects, extreme events, and unwanted side effects", implying that the potential for cascading disruption is a growing and critical concern for many facets of daily life'.

Linkov et al. [21] reports that 'how such interconnectivity facilitates stochastic, non-linear, and spatially interspersed causal structures that, if triggered, contribute to a 'domino effect' that can permanently alter a broader infrastructural, environmental, or social system'. Inherent vulnerabilities within societal systems such as that exhibited by poverty, inequality, poorly maintained critical infrastructure become apparent when impacted by a triggering event.

Sensemaking thereby becomes a key factor in revealing not only threats and risk but also the vulnerabilities that permeate the societal system as 'resident pathogens' [41].

The impacts of disaster events cannot be seen in isolation. WEF Global Risks reports have clearly articulated the hyperconnected risks that permeate the world and reflect the inherent '...fragility and vulnerabilities that lie within the social/technological/economic/political/ecological interdependent systems' [30]. It is through these underlying networks that Helbing [10] argues that we have '…created pathways along which dangerous and damaging events can spread rapidly and globally' and thereby has increased systemic risks. A Chatham House report, 'Preparing for High Impact, Low Probability Events', found that governments and businesses remain unprepared for such events (Lee et al. [20]. Helbing [10] argues that:

> Many disasters in anthropogenic systems should not be seen as 'bad luck', but as the results of inappropriate interactions and institutional settings. Even worse, they are often the consequences of a wrong understanding due to the counter-intuitive nature of the underlying system behaviour.

This chapter will explore the impact of Hurricane Maria on Puerto Rico and discuss the vulnerabilities that were inherent in the societal system. Sensemaking will be used as a lens in the analysis that will point to the requirement for disaster risk reduction and community resilience in societal systems. Enabling resilience and building back better have more to do than just with the immediate impacts of hurricanes but also include understanding the secondary and tertiary impacts arising from inherent vulnerabilities in societal systems.

2 Sensemaking to Support Vulnerability Analysis

The concept of 'sensemaking' [67] is an essential element to understanding disaster aetiology and inherent vulnerabilities in a societal system. Weick [67] refers to sensemaking in terms of '…how we structure the unknown so as to be able to act in it. Sensemaking involves coming up with a plausible understanding—a map—of a shifting world; testing this map with others through data collection, action, and conversation; and then refining, or abandoning, the map depending on how credible it is' (cited in [1]).

As noted in (McCrae [23]:2), the failure of sensemaking is strongly implicated in many studies of organizational accidents and disaster (e.g. [52, 57, 66]). As such, organizational sensemaking offers a valuable set of ideas that are well-suited to studying the early processes of noticing, identifying and interpreting signs of unknown, latent risks in organizations'. The aetiology of disasters is often paved by actions and decisions preceding the event. These actions and decisions are rooted in mental models, beliefs and assumptions that are often volatile and vulnerable to trigger events that reveal disfunctionalities. This is captured well in the work of [57–59]; [23, 26, 63, 64].

Systems thinking figures prominently in the discourse pertaining to disaster aetiology and disaster forensics [29]. Jackson [12] defines systems thinking paradigm as'…a discipline for seeing the 'structures' that underlie complex situations, and for discerning high from low leverage change…Ultimately, it simplifies life by helping us to see the deeper patterns lying beneath the events and the details'. In this complex problem landscape of systemic risks, trigger events and cascading public health issues, systems thinking emerges as both a worldview and a process in the sense that it informs ones understanding regarding a system and can be used as an approach in problem solving [7]. Peters [39] argues that '…at its core, systems thinking is an enterprise aimed at seeing how things are connected to each other within some notion of a whole entity. We often make connections when conducting and interpreting research, or in our professional practice when we make an intervention with an expectation of a result. Anytime we talk about how some event will turn out, whether the event is an epidemic, a war, or other social, biological, or physical process, we are invoking some mental model about how things fit together. However, rather than relying on implicit models, with hidden assumptions and no clear link to data, systems thinking deploys explicit models, with assumptions laid out that can be calibrated to data and repeated by others'.

Systems thinking thereby is a key sensemaking approach to view inherent vulnerabilities in societal systems that lie dormant, waiting for a trigger event to unleash a cascading disaster across communities.

3 Case Study: Hurricane Maria and Puerto Rico

The 2017 North Atlantic Hurricane Season, June 1st through November 30th, had one of the highest counts of named tropical storms and hurricanes on record since 1851. Specifically, the month of September 2017 had the most hurricane days every recorded [15]. Not only was the number of storms that season noteworthy, but the damage was also considerable. For the three hurricanes that caused the most damage to the United States (US), Hurricane Harvey, Hurricane Irma, and Hurricane Maria, Congress approved a total of $33.92 billion dollars in disaster funding [65].

Hurricane Maria hit the United States territory of Puerto Rico on September 20th, 2017, as a category four storm. It's estimated that thousands in Puerto Rico died due to the storm, and in many ways, the island was devastated from the disaster. For instance, it caused the longest blackout ever in the US, which likely contributed to the mortality count, roads were either washed away or severely damaged, wireless communications were inoperable, and water treatment facilities weren't functioning to provide potable water [9, 46]. The federal government mounted a large-scale disaster response to mitigate damage but even three years after the storm Puerto Rico still had not recovered to what it was pre-Maria [9, 46].

To grasp the extent of the damage from Hurricane Maria, it's best to first understand in what ways Puerto Rico was vulnerable prior to it. Economically, Puerto Rico is one of the poorest areas in the US, with 44.9% of people (41.2% of households) living below the Federal Poverty Level (FPL) in 2017 based a five-year estimate from 2013 data [60]. From those same estimates, 61.4% of Puerto Rican residents rely on public health insurance, and nearly half of the population in a municipio (municipality, i.e. county) designated as a health care professional shortage area [60]. Moreover, due to economic downturn over the 10 years before Hurricane Maria, an estimated 12% of Puerto Rico's residents had moved to a US state resulting in the demographic composition of Puerto Rico being proportionally older than other US states. Specially, 18% of Puerto Ricans being over the age of 65 [46].

The severity of Hurricane Maria's long-lasting impact on Puerto Rico is a clear example of a community that had limited resources to brace against a catastrophe or rebound after it.

4 Discussion

4.1 Impacts of Hurricane Maria

Power Outage

Hurricane Maria caused severe damage to Puerto Rico's electric and alternative energy infrastructure and resulted in over four billion customer hours of power outage; the longest in US history [45]. Prolonged power outages have cascading effects in the modern world that impact far more than the comforts of air conditioning and television. They easily can, and in post-Hurricane Maria Puerto Rico did, lead to public health impacts such as strains on the care hospitals and emergency departments can provide, harming those that have a medical condition that leaves them dependent on electricity or refrigeration of medication, heat stroke, and barriers to communication if emergency services are needed [9, 14].

At the time that Hurricane Maria hit Puerto Rico, the infrastructure of its electrical system was already in a vulnerable state. Puerto Rico's Electric Power Authority (PREPA), the only utility supply company to Puerto Rico, was $9 billion dollar in debt with reports of major cuts to their workforce and the power grid not being well maintained [9]. Making matters worse, PREPA did not request assistance with restoration for over a month after the storm and instead negotiated two no-competition contracts with small, inexperienced companies with the stipulation that those companies not be audited [9].

The Army Corps of Engineers spearheaded the system's restoration and eventually the American Public Power Association provided mutual aid and the PREPA CEO resigned [9]. However, even with this aid, older equipment, mountainous terrain, and the constraint of resources having to be shipped in from the mainland were immense barriers to restoring the system [9]. Furthermore, the prolonged power outages caused ripple effects to other sectors such as government laboratory testing for disease surveillance, pharmacies filling prescription medication, and critical information being disseminated to the public by the news [4, 34, 51]. Though most of Puerto has had electricity restored, there are still reports of remote areas that have not been restored and reoccurring outages (Ortiz-Blanes [38]).

Years of poor maintenance and the impact from Hurricane Maria resulted in extensive damage to Puerto Rico's electric and power infrastructure that caused disruptions for months to millions of customers and other sectors that the public depend on. Uncertainty of leadership, pressure on the speed of restoration, the difficult terrain and isolated location of Puerto Rico, and the poor condition of existing equipment have made repairing and rebuilding a challenge. Puerto Rico would require a large amount of resources and time to establish an electric and power infrastructure that could withstand another hurricane.

During the recovery period after Hurricane Maria, one of the first priorities was quickly restoring power. However, the other competing priorities for this sector were considered such as focusing on resilience to withstand another hurricane, stability to reduce reoccurring outages, and being economically sound to not incur more debt. One option for recovery was to replace most, if not all, of the infrastructure with underground power lines. This option provides most resilience and sustainability because underground lines are less likely to be impacted by a hurricane and a new electric grid would mean that it wasn't relying on older, less stable equipment. However, this would have cost an estimated $60 billion dollars and likely take longer to implement than the public could wait for [9]. Another option would have also replaced the existing infrastructure, but with renewable energy such as solar, wind, and hydro powered generators. This would eventually be economical but only after the costly installation and would be sustainable except for natural disasters such as hurricanes and floods. The few solar and wind renewable energy sources that were available on Puerto Rico prior to Hurricane Maria were destroyed by the storm, which showed its limitations. However, some customers switched to relying on solar power for a few vital resources in the months that they waited for power to be restored [9, 17]. Ultimately, to restore power as quickly as possible to the public, it was decided that the existing infrastructure would be repaired, as much equipment as possible that was available would be used, and equipment would be replaced only when absolutely necessary. However, this has shown to be shortsighted because Puerto Rico still experiences outages and the system is unlikely to be able to withstand another major hurricane ([9], Ortiz-Blanes [38]).

Mortality

Months after Hurricane Maria, the official fatality count in Puerto Rico remained at 64 despite growing public opinion that this was a gross underestimation. Delays in reporting these deaths were multi-faceted and complex. Firstly, a death could not be determined to be a direct result of the storm unless it was transported to San Juan, PR and confirmed by the Institute of Forensic Sciences (United States Government Accountability Office [USGAO] [61]). The only alternative being if a medical examiner traveled to the municipality where the body was held to confirm the cause of death. Secondly, this did not account for indirect disaster-related deaths such as exacerbation of chronic conditions, delays in receiving medical attention due to infrastructure damage, or increases in suicides or homicides. Several studies have attempted to statistically analyze the fatality count in Puerto Rico due Hurricane Maria.

The fatality count of a disaster is one way of measuring the impact that an event had on an area. However, that count should not only include those who died as a direct result of a disaster, but also those lives lost after the event and who wouldn't have died if the event hadn't happened. For example, an individual who drowns during a hurricane is a direct death and an individual who dies from a heart attack at home five days after that hurricane because an ambulance cannot reach them for many hours is an indirect death. If not for the hurricane, both individuals in all probability might not have died.

Additionally, in recent years statistical analysis of an area's excess mortality after an event has uncovered an even more informative method for measuring the impact of a disaster. Excess mortality is the significant increase of all mortality in an area for prolonged periods after an event and refers to ripple effects of a disaster beyond the scope of indirect deaths. It's important to note that there is no set standard for counting fatality disasters, whether they're direct, indirect, or excess mortality.

Seven studies have been published from research groups that endeavored to enumerate the Puerto Rican lives lost due to Hurricane Maria; each with their own unique methodology. One of the first of these studies was disseminated in July 2018 and took the approach of surveying a randomized sample of households. The survey was conducted in early 2018 and included questions on those who had died in 2017 and if post-storm what was their cause of death, those that had moved out of the area or had not been seen since Hurricane Maria, and access to electricity, water, and cell phone service [14]. This unique approach has been previously used to determine excess mortality from war (Spagat and van Weezel [53]). Kishore et al. estimated that 4,645 (95% CI, 793—8498) fatalities between September 20, 2017 and December 21, 2017, the highest estimation of all the studies and the study with the widest confidence interval range [14].

The territory sponsored a study that partnered George Washington University and the University of Puerto Rico to provide an accurate estimate of how many lives lost, either directly or indirectly, due to Hurricane Maria and the demographics of those excess fatalities. Their statistical analysis found that between September 2017 and February 2018 that 2,975 (95% CI, 2658–3290) more individuals died than expected based on mortality trends from the previous years [47]. Among these excess fatalities, higher rates were seen in men, those aged 40 and old, and those living in municipalities with lower socioeconomic status than trends from previous years [47]. The methodology of this study was unique to others because it adjusted for post-Maria displacement and it found significant mortality increases for six months post-Maria [47].

Later similar studies of excess mortality in Puerto Rico after Hurricane Maria conducted additional statistical sensitivity testing, found that less months post-Maria had significant increases in mortalities, and justified not adjusting for displacement. Though these studies differ in their methodology, they all agree that in September and October 2017 there was an increase in mortalities in Puerto Rico that can be contributed to Hurricane Maria and their mortality estimates capture a more comprehensive view of the impact it caused to the health of Puerto Ricans. When looking at this shorter 2-month time period of increased fatalities, all studies find that roughly 1,000 excess mortalities occurred [14]; Rivera and Rolke [42, 43]; Santos-Burgoa et al. [47]; Santos-Lozada and Howard [48]; Spagat and van Weezel [53]; plus 5 main studies). These studies led to the local government of Puerto Rico rejecting the initial fatality count of 64 and officially accepted 2,975 as the fatality count in Puerto

Rico due to Hurricane Maria. Despite this being the highest hurricane fatality count in the U.S., there is evidence that disaster aid and resources were proportionately far less for Hurricane Maria than Hurricanes Harvey and Irma that also impacted the US in 2017 [65].

Long-Term Impacts

In addition to the fatalities and severe damages caused by Hurricane Maria, Puerto Rico also experienced a prolonged period of recovery and redevelopment. Even years later, some areas and sectors still aren't fully recovered and others that were already struggling pre-Maria are now worse than before.

It's well documented that disasters have long-term effects on the mental health of the affected population with many experiencing post-traumatic stress disorder (PTSD), anxiety, depression, and grief [3]. Because of their pivotal, first-hand, role in disaster response and recovery, essential workers, such as health care and social service providers, are markedly susceptible to these mental health symptoms and burnout without support and resilience [40]. Essential workers in Puerto Rico faced months of increased demand for essential health services after Hurricane Maria and it was found that they reported higher amounts of emotional distress than their counterparts in Texas after Hurricane Harvey [40]. While workers in both locations reported having social support, those in Puerto Rico reported having lower resilience which could have contributed to their higher mental health symptoms [40].

In the US after Hurricanes Andrew and Katrina, studies showed that the mental health of children was particularly impacted due to the chaos of changes in daily routine to school and community activities (La Greca et al. [16, 18]). Puerto Rico was no exception to this with studies finding that children were still experiencing signs of PTSD and depression five to nine months after Hurricane Maria [37]. In some respects, children in Puerto Rico were considered a vulnerable population before Hurricane Maria because many more of them, 58%, live below the Federal Poverty Level (PFL) than Puerto Rican-origin children in other parts of the US, 32%, or the overall US child poverty rate of 20% [31, 50].

As expected with a natural disaster, Puerto Rico has also faced long-lasting envi-ronmental impacts due to Hurricane Maria and the ecology could take years to recover. Forests, agriculture, and native species of animals and insects were affected worse than damage from previous hurricanes. For example, an ecology survey of tree damage and destruction found that twice as many trees were impacted during Hurricane Maria than when Hurricane Hugo hit Puerto Rico in 1989 [62]. Though tree damage may not initially seem worth mentioning compared to other degrees of damage from a hurricane, it is an indicator of damage to roadways, powerlines, and housing. All of which could lead to injuries, fatalities as well as a decrease in tourism and reveals years or decades of damage to ecology systems.

Puerto Rico suffered $780 million in agricultural damages from Hurricane Maria. Initially it was estimated that 90% of the local coffee plants, a major crop for many Caribbean islands, were destroyed though later surveys assessed that the damage wasn't as extensive [11, 13, 25, 33]. It was hoped that a silver lining to the hurricane damage would be that the invasive insect known as the coffee berry borer (Hypothen-emus hampei) would be wiped out or at least decreased. The species was first found in Puerto Rico in 2007 and it has severely harmed the coffee crops of Puerto Rico ever since [32]; Marino et al. [24]. Unfortunately, though much of the coffee berry borer population was blown away by Hurricane Maria, its eggs survived in coffee berries and quickly repopulated.

The two years prior to Hurricane Maria, Puerto Rico experienced an outbreak of Zika, a mosquito-borne virus that for exposed pregnant women can lead to congenital abnormalities or fetal loss. Puerto Rico also is also endemic to other serious and sometimes fatal arboviruses such as dengue and chikungunya, which are also spread by the same mosquito species as Zika, Aedes aegypti. Though a hurricane will first blow away local mosquitoes, shortly after there can be a surge because hurricanes bring flooding, rain, and debris to the areas they affect which lead to an increase in potential mosquito breeding locations [2]. Naturally, post-Maria Puerto Rico was concerned about a rise in arboviruses that could come with this rise in mosquito breeding. Though there was a sharp increase in the dreaded Aedes aegypti five weeks post-Maria, fortunately mosquito testing found them not to be carrying Zika, dengue, or chikungunya viruses and there was no increase in human cases [2].

By the time Hurricane Maria hit, the economy of Puerto Rico had already been in a decline for over 10 years largely due to manufacturing companies leaving the territory after the expiration of once attractive federal tax incentives (Llorens [22]; Santos-Lozada et al. [49]). This also led to Puerto Rico experiencing a population decline from both emigration and naturally from more deaths than births (Rodriguez-Diaz [44]; Santos-Lozada et al. [49]). Though Puerto Rico was experiencing these setbacks prior to Hurricane Maria, it appears as though it only exacerbated the situation. Post-Maria there was a large spike of outbound traffic to the US mainland for evacuees (Santos-Lozada et al. [49]). They were displaced across 37 states and some weren't able to return for over six months while other chose to permanently relocate (Torrens et al. [56]). Studies have shown that this outflux was temporary and many Puerto Rican residents did eventually return, though the steady population decline still continued. However, there is also evidence that the declining employment rate of Puerto Rico also experienced a post-Maria spike. Unemployment numbers began to recover months afterward though they remain worse than pre-Maria and it's possible that the COVID-19 pandemic has set off another spike that it can't recover from (Santos-Lozada et al. [49]).

5 Sensemaking Vulnerability and Risk

One of the key challenges facing disaster risk managers and community planners is the identification of vulnerabilities and unforeseen risks that reside within the physical, human and informational domains of societal critical infrastructure. Latent risks [41] and vulnerabilities seeded into societal critical infrastructure [27] point to the requirement for sensemaking that makes them visible.

Sensemaking in crisis and disaster management operationally means, as articulated in Nowling and Seeger [35], '...they scan their environments for cues of changes and operational threats. After identifying the threat or change, they communicate and work collaboratively to determine the most plausible meaning of these cues [66]. They then enact decisions and actions designed to reduce risk and resolve organizational challenges. When this process is deficient, Weick adds, even the most well-managed organizations can escalate their exposure to risks of crisis'. Such sensemaking practices reveals opportunities for disaster risk reduction and resilience building. Resilience can be construed as a 'global state of preparedness where targeted systems can absorb unexpected and potentially high consequence shocks and stresses [19], [21]. As noted in the case study, the systems lens is essential in understanding the connectivity and interdependencies in societal critical infrastructure, latent risks and the cascading events that may be triggered by a disaster event. Resilience affords greater clarity over such threats (particularly systemic threats) by focusing upon the inherent structure of the system, its core characteristics, and the relationship that various sub-systems have with one another to generate an ecosystem's baseline state of health (IRGC [8], [21]).

Sensemaking applying systems thinking approaches, reveal how the structure and activities of systems influence one another, and serves as an avenue to understand and even quantify a web of complex interconnected networks and their potential for disruption via cascading systemic threat [21].

Although a disaster may be triggered by a natural hazard, its effect on society is grounded in the social system in which it takes place. As the case study on Hurricane Maria shows, community vulnerability is constructed over many years and is influenced by social/political/economic factors. Understanding the complexity of vulnerability to disaster events is at the heart of disaster risk reduction and resilience building. Hence sensemaking recognizes the importance of understanding the impact of mental models, beliefs, assumptions and agendas on resilience.

Sensemaking thereby emerges as a key element in understanding disaster, risks and resilience with a focus on the surrounding social environments in which vulnerability exist. The case study revealed how vulnerability varies across physical space and across social groups. These vulnerabilities reside within the societal infrastructure and often emerge as public health disasters triggered by a hazard event (Hurricane). Within the context of disaster events and the Hazard of Place model [6], Hurricane Maria impact on Puerto Rico shows how societal infrastructure vulnerabilities are reflected in the physical and social dimensions (Fig. 1).

The analysis presented along with the lens of the Hazard of Place model illuminates opportunities and leverage points for building resilience and disaster risk reduction. As noted in (OECD [36]:12–13) 'Recovery and adaptation in the aftermath of disruptions is a requirement for interconnected twenty-first Century economic, industrial, social, and health-based systems, and resilience is an increasingly crucial part of strategies to avoid systemic collapse. Based on NAEC reports and the resilience literature, specific recommendations for building resilience to contain epidemics and other systemic threats include:

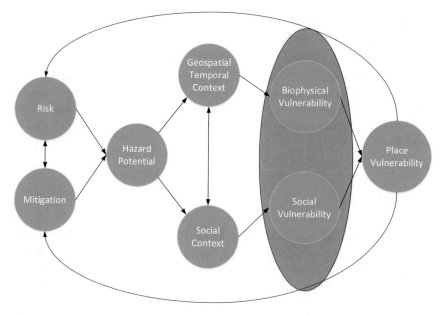

Fig. 1 Hazard of Place model [6]

1. Design systems, including infrastructure, supply chains, and economic, financial and public health systems, to be resilient, i.e. recoverable and adaptable.
2. Develop methods for quantifying resilience so that trade-offs between a system's efficiency and resilience can be made explicit and guide investments.
3. Control system complexity to minimise cascading failures resulting from unexpected disruption by decoupling unnecessary connections across infrastructure and make necessary connections controllable and visible.
4. Manage system topology by designing appropriate connection and communications across interconnected infrastructure.
5. Add resources and redundancies in system-crucial components to ensure functionality.
6. Develop real-time decision support tools integrating data and automating selection of management alternatives based on explicit policy trade-offs in real time'.

The IRGC highlights a multi-step procedure to identify, analyse, and govern systemic risks, as well as better prepare affected systems for such risks by mitigating possible threats and transitioning the system towards one of resiliency-by-design. The IRGC's cyclical process for the governance of systemic risk includes:

7. Explore the system, define its boundaries and dynamics.
8. Develop scenarios considering possible ongoing and future transitions.
9. Determine goals and the level of tolerability for risk and uncertainty.
10. Co-develop management strategies dealing with each scenario.

11. Address unanticipated barriers and sudden critical shifts.
12. Decide, test and implement strategies.
13. Monitor, learn from, review and adapt (OECD [36]:12–13).

Sensemaking is critical in vulnerability analysis to support disaster risk reduction and resilience. As noted in the case study, latent risk stemming from decisions and actions displaced in time and space from the trigger event, created inherent vulnerabilities that precipitated a cascading effect through the societal critical infrastructure including public health.

6 Conclusion

Hurricane Maria devastated Puerto Rico. Puerto Rico's higher poverty levels, aging population, and struggle to provide social services have made the effects of Hurricane Maria that much more profound for its residents. Coupled with the aging and poorly maintained societal infrastructure, the trigger event of the hurricane revealed 'latent risks' within the societal systems. With consideration of the Hazard Place model, Taleb [55] comments certainly resonate:

> Not seeing a tsunami or an economic event coming is excusable; building something fragile to them is not.

Given the case study results pointing to cascading effects triggered by the hurricane across the vulnerabilities within the societal infrastructure, sensemaking applying the systems thinking approach, as noted in the NAEC report (OECD [36]) provides an avenue to support disaster risk reduction and resilience (the ability of a system to anticipate, absorb, recover from, and adapt to a wide array of systemic threats) (OECD [36]:11).

References

1. Ancona DL (2011) SENSEMAKING framing and acting in the unknown in handbook of teaching leadership, 3–20. Sage Publications Inc, Thousand Oaks, CA
2. Barrera R, Felix G, Acevedo V, Amador M, Rodriguez M, Rodriguez D, Rivera L, Gonzalez O, Nazario N, Ortiz M, Munoz-Jordan J, Waterman S, Hemme RR (2019) Am J Trop Med Hyg 100(6). https://doi.org/10.4269/ajtmh.19-0015
3. Bonanno GA, Brewin CR, Kaniasty K, La Greca AM (2010) Weighting the costs of disasters: consequences, risks, and resilience in individuals, families, and communities. Psychol Sci Public Interest 1(1). https://doi.org/10.1177/1529100610387086
4. Concepcion-Acevedo J, Patel A, Luna-Pinto C, Pens RG, Ruiz RIC, Arbolay HR, O'Neill E et al (2018) Initial public health laboratory response after hurricane maria—puerto rico, 2017. Morb Mortal Wkly Rep 67(11):333–336
5. Centeno MA, Nag M, Patterson TS, Shaver A, Windawi AJ (2015) The emergence of global systemic risk. Ann Rev Sociol 41:65–85
6. Cutter SL (1996) Vulnerability to environmental hazards. Progress Hum Geogr 20:529–539

7. Edson R (2008) Systems thinking. Applied: a primer. ASysT Institute, http://www.asysti.org/OurModules/SharedDocs/HistoryDocView.aspx?ID=373
8. IRGC (2018) Trump BD, Florin MV, Linkov I (2018) IRGC resource guide on resilience, vol 2, (No. BOOK). International Risk Governance Center (IRGC)
9. Gallucci M (2018) Rebuilding puerto rico's grid eight months after hurricane maria, electricity is nearly restored-but that's just the beginning. *Ieee Spectrum, 55*(5), 30–38. Retrieved from < Go to ISI > ://WOS:000431446400014
10. Helbing D (2013) Globally networked risks and how to respond. Nature 497:51–59
11. Hoffman M (2017) Hurricane maria devastated puerto rico's coffee farms. Oct. 11, 2017. Retrieved from: https://www.eater.com/2017/10/11/16434030/puerto-rico-hurricane-maria-coffee-farms-destroyed-how-to-help. Accessed 1 Aug 2018
12. Jackson MC (2003) Systems thinking: creative holism for managers. Wiley, West Sussex
13. Kennedy M (2017) 'This was a beautiful place', Puerto Rico's coffee farms devastated by Maria. Oct. 10, 2017. Retrieved from: https://www.npr.org/sections/thetwo-way/2017/10/10/556657967/-this-was-a-beautiful-place-puerto-rico-s-coffee-industry-devastated-by-maria. Accessed 1 Aug 2018
14. Kishore N, Marques D, Mahmud A, Kiang MV, Rodriguez I, Fuller A, Buckee CO (2018) Mortality in Puerto Rico after hurricane maria. New England J Med 379(2):162–170. https://doi.org/10.1056/NEJMsa1803972
15. Klotzbach PJ, Schreck CJ, Collins JM, Bell MM, Blake ES, Roache D (2018) The extremely active 2017 North Atlantic hurricane season. Monthly Weather Rev 146(10):3425–3443. https://doi.org/10.1175/mwr-d-18-0078.1
16. Kroneberg ME, Hansel TC, Brennan AM, Osofsky HJ, Osofsky JD (2010) Children of katrina: lessons learned about postdisaster symptoms and recovery patterns. Child Dev 81(4)
17. Kwasinski A (2018) Effects of hurricane maria on renewable energy systems in puerto rico. In: 2018 7th international conference on renewable energy research and applications, pp 383–390
18. La Greca AM, Silverman WK, Vernberg EM, Prinstein MJ Symptoms of posttraumatic stress in children after Hurricane Andrew; a prospective study. J Consult Clin Psychol 64(4)
19. Larkin S, Fox-Lent C, Eisenberg DA, Trump BD, Wallace S, Chadderton C, Linkov I (2015) Benchmarking agency and organisational practices in resilience decision making. Environ Syst Decisions 35(2):185–195
20. Lee B, Preston F, Green G (2012) Preparing for high-impact, low-probability events lessons from eyjafjallajökull- a chatham house report. https://www.chathamhouse.org/sites/default/files/public/Research/Energy,%20Environment%20and%20Development/r0112_highimpact.pdf
21. Linkov I, Trump BD, Hynes W (2019) Resilience-based strategies and policies to address systemic risks. SG/NAEC(2019)5
22. Lloréns H (2018) Ruin nation. NACLA Rep Am 50(2):154–159. https://doi.org/10.1080/10714839.2018.1479468
23. Macrae C (2007) Interrogating the unknown: risk analysis and sensemaking in airline safety oversight LSE DISCUSSION PAPER NO: 43 DATE: May 2007 http://eprints.lse.ac.uk/36123/1/Disspaper43.pdf
24. Mariño YA, Vega VJ, García JM, Rodrigues V, José C, García NM, Bayman P (2017) The coffee berry borer (Coleoptera: Curculionidae) in puerto rico: distribution, infestation, and population per fruit. J Insect Sci 17
25. Marino YA, Garcia-Pena JM, Vega VJ, Verle-Rodrigues JC, Bayman P (2018) After the fall: did coffee plants in Puerto Rico survive the 2017 hurricanes? Agricult Ecosyst Environ 266:10–16. https://doi.org/10.1016/j.agee.2018.07.011
26. Masys AJ (2010) Fratricide in air operations: opening the black box—revealing the social. Ph.D. Dissertation, June 2010, University of Leicester, UK
27. Masys AJ (ed) (2014) Networks and network analysis for Defence and Security. Springer Publishing
28. Masys AJ (ed) (2015) Disaster management: enabling resilience. Springer Publishing

29. Masys AJ (ed) (2016) Disaster forensics: understanding root cause and complex causality. Springer Publishing
30. Masys AJ, Ray-Bennett N, Shiroshita H, Jackson P (2014) High impact/ low frequency extreme events: enabling reflection and resilience in a hyper-connected world. 4th international conference on building resilience, 8–11 September 2014, Salford Quays, United Kingdom. Procedia Economics and Finance 18(2014):772–779
31. Mayol-Garcia YH (2020) Pre-hurricane linkages between poverty, families, and migration among Puerto Rican-origin children living in Puerto Rico and the United States. Population Environ 42. https://doi.org/10.1007/s11111-020-00353-7
32. NAPPO (2007) Detection of coffee berry borer, hypothenemus hampei, in Puerto Rico—United States. Official Pest Reports
33. Newton T, Quiñones-García KL (2017) La industria de café en Puerto Rico: después de Huracán María. Nov. 8. Retrieved from: https://www.perfectdailygrind.com/2017/11/la-industria-de-cafe-de-puerto-rico-despues-de-huracan-maria/. Accessed 1 Aug 2018
34. Nieves-Pizarro Y, Takahashi B, Chavez M (2019) When everything else fails: radio Journalism During Hurricane Maria in Puerto Rico. Journalism Practice 13(7):799–816. https://doi.org/10.1080/17512786.2019.1567272
35. Nowling WD, Seeger MW (2020) Sensemaking and crisis revisited: the failure of sensemaking during the Flint water crisis. J Appl Commun Res 48(2):270–289
36. OECD Policy Responses to Coronavirus (COVID-19) A systemic resilience approach to dealing with Covid-19 and future shocks 28 April 2020. http://www.oecd.org/coronavirus/policy-responses/a-systemic-resilience-approach-to-dealing-with-covid-19-and-future-shocks-36a5bdfb/
37. Orengo-Aguayo R, Stewart RW, de Arellano MA, Suarez-Kindy JL, Young J (2019) Disaster exposure and mental health among puerto rican youths after hurricane maria. Jama Network Open 2(4): https://doi.org/10.1001/jamanetworkopen.2019.2619
38. Ortiz-Blanes S (29 July 2020) Puerto Rico's power grid fails hours ahead of potential arrival of tropical storm. The Miami Herald. https://www.miamiherald.com/news/nationworld/world/americas/article244571552.html
39. Peters DH (2014) The application of systems thinking in health: why use systems thinking? Health Res Policy Sys 12:51
40. Powell TM, Yuma PJ, Scott J, Suarez A, Morales I, Vinton M, Li S.-J et al (2019) In the aftermath: the effects of hurricanes harvey and maria on the well-being of health-care and social service providers. Traumatology. https://doi.org/10.1037/trm0000228
41. Reason J (1997) Managing the risk of organizational accidents. Ashgate Publishing, Aldershot, England
42. Rivera R, Rolke W (2018) Estimating the death toll of Hurricane María. Significance 15(1):8–9
43. Robles F, Davis K, Fink S, Almukhtar S (2017) Official toll in Puerto Rico: 64. Actual deaths may be 1,052. New York Times 8
44. Rodríguez-Díaz CE (2018) Maria in Puerto Rico: natural disaster in a colonial archipelago. Am J Public Health. https://doi.org/10.2105/AJPH.2017.304198
45. Roman MO, Stokes EC, Shrestha R, Wang ZS, Schultz L, Carlo EAS, Enenkel M et al (2019). Satellite-based assessment of electricity restoration efforts in Puerto Rico after hurricane maria. PloS One 14(6). https://doi.org/10.1371/journal.pone.0218883
46. Rudner N (2019) Disaster care and socioeconomic vulnerability in Puerto Rico. J Health Care Poor Underserved 30(2):495–501. https://doi.org/10.1353/hpu.2019.0043
47. Santos-Burgoa C, Sandberg J, Suarez E, Goldman-Hawes A, Zeger S, Garcia-Meza A, Goldman L et al (2018) Differential and persistent risk of excess mortality from hurricane maria in Puerto Rico: a time-series analysis. Lancet Planet Health 2(11). https://doi.org/10.1016/s2542-5196(18)30209-2
48. Santos-Lozada AR, Howard JT (2018) Use of death counts from vital statistics to calculate excess deaths in Puerto Rico following hurricane maria. J Am Med Assoc 320(14). https://doi.org/10.1001/jama.2018.10929

49. Santos-Lazoda AR, Kaneshrio M, McCarter C, Marazzi-Santiago M (2020) Puerto Rico exodus: long-term economic headwinds prove stronger than hurricane maria. Popul Environ 42(43). https://doi.org/10.1007/s11111-020-00355-5
50. Semega JL, Fontenot KR, Kollar MA (2017) Income and poverty in the United States: 2016. U.S. Census Bureau, Current Population Reports P60–259. Accessed on 8/13/2018. Available at https://www.census.gov/content/dam/Census/library/publications/2017/demo/P60-259.pdf
51. Smith JY, Sow MM (2019) Access To E-prescriptions and related technologies before and after hurricanes harvey, irma and maria. Health Aff 38(2):205–211. https://doi.org/10.1377/hlthaff.2018.05247
52. Snook SA (2000) Friendly fire: the accidental shootdown of US Black Hawks over Northern Iraq. Princeton University Press, Oxford
53. Spagat M, van Weezel S (2020) Excess deaths and hurricane maria. Popul Environ 42. https://doi.org/10.1007/s11111-020-00341-x
54. Taleb NN (2007) The black swan: the impact of the highly improbable. Penguin Books Ltd, London
55. Taleb NT (2014) Antifragile: things that gain from disorder. Random House, New York
56. Torrens C, Saloman G, Coto D (7 Mar 2018) Puerto ricans still stranded in hotels 6 months after storm. Associated Press. https://apnews.com/article/d90082fdcb6443fda2ac42c32524a020
57. Turner B (1976) The organizational and interorganizational development of disasters. Adm Sci Q 21:378–397
58. Turner B (1978) Man-made disasters. Wykeham, London
59. Turner B, Pidgeon N (1997) Man-made disasters, 2nd edn. Butterworth-Heinemann, Oxford
60. United States Census Bureau (2019) The American community survey. Retrieved from https://factfinder.census.gov/faces/nav/jsf/pages/searchresults.xhtml?refresh=t
61. United States Government Accountability Office (2019) Disaster response: federal assistance and selected states and territory efforts to identify deaths from 2017 hurricanes. https://www.gao.gov/reports/GAO-19-486/
62. Uriarte M, Thompson J, Zimmerman JK (2019) Hurricane Maria tripled stem breaks and doubled tree mortality relative to other major storms. Nat Commun 10(1):1362. https://doi.org/10.1038/s41467-019-09319-2
63. Vaughan D (1990) Autonomy, interdependence, and social control: NASA and the space shuttle challenger. Adm Sci Q 35:225–257
64. Vaughan D (1996) The challenger launch decision: Risky technology, culture and deviance at NASA. Chicago University Press, London
65. Willison CE, Singer PM, Creary MS, Greer SL (2019) Quantifying inequities in US federal response to hurricane disaster in Texas and Florida compared with Puerto Rico. Bmj Global Health 4(1). https://doi.org/10.1136/bmjgh-2018-001191
66. Weick KE (1993) The collapse of sensemaking in organizations: The Mann Gulch disaster. Adm Soc 38(4):628–652
67. Weick KE (1995) Sensemaking in organizations. Sage, Thousand Oaks, CA
68. Weick KE, Sutcliffe KM (2007) Managing the unexpected: resilient performance in an age of uncertainty, 2nd edn. Wiley, San Francisco

A System Dynamics Model of COVID-19 in Canada: A Case Study in Sensemaking

Ivan Taylor and Anthony J. Masys

Abstract The world is becoming increasingly vulnerable to infectious diseases, creating a global health security issue. Over the last 2 decades, many global and national health crises have emerged such as SARS, H5N1, H1N1, and now COVID-19. The recent COVID-19 pandemic reflects how unexpected events often audit our resilience (Weick and Sutclifffe [10]. The mortality and morbidity statistics associated with COVID-19 has become a key impact metric. At the time of publication, in Canada, upwards of 675,000 cases of COVID 19 have been reported and 17,500 deaths (https://health-infobase.canada.ca/covid-19/epidemiological-summary-covid-19-cases.html?stat=num&measure=total&map=pt#a2). The pandemic has tested and left wanting the global ability to respond to such a threat. Heyman et al. [3] argue that "the world is ill-prepared" to handle any "sustained and threatening public-health emergency." Such public health emergencies stemming from infectious disease outbreaks are creating a serious threat to societal well-being and national security. The inherent interconnectivity and interdependency within societal public health systems require analysis that provides a deep understanding regarding the potential impact of COVID-19 on populations in response to intervention strategies. In dynamic systems, the effects of an intervention are only evident after a time delay. Understanding the system and its inherent dynamics is a key requirement for sensemaking and is a game-changer in supporting crisis management of complex issues such as a global pandemic. This chapter examines a case study of early sensemaking about the COVD-19 pandemic in Canada through the application of System Dynamics.

Keywords System dynamics · COVID-19 · Systems thinking · Pandemic · Sensemaking

I. Taylor
Policy Dynamics, New Hamburg, Canada

A. J. Masys (✉)
College of Public Health, University of South Florida, Tampa, USA

International Centre for Policing and Security, University of South Wales, Newport, Wales, UK

1 Introduction

The COVID 19 pandemic has resulted in a significant national and global public health crisis. The mortality and morbidity statistics associated with COVID-19 has become a key impact metric. At the time of publication, in Canada, upwards of 675,500 cases of COVID 19 have been reported and 17,500 deaths.[1] Within Canada, the Covid-19 outbreak has resulted in a considerable impact on public health security, human security, and has also impacted the national economy. As described in OECD [7] 'the pandemic has reminded us bluntly of the fragility of some of our most basic human-made systems. Shortages of masks, tests, ventilators, and other essential items have left frontline workers and the general population dangerously exposed to the disease itself. At a wider level, we have witnessed the cascading collapse of the entire production, financial, and transportation systems, due to a vicious combination of supply and demand shocks'.

As cited in Linkov et al. [6], 'Helbing [2] notes that the consequences of failing to appreciate and manage the characteristics of complex global systems and problems can be immense'. This is equally applicable to national societal systems as well. Developing public health intervention strategies associated with the COVID-19 pandemic thereby requires sensitivity to unexpected outcomes, to sensemaking that explores the problem space from a systems perspective. As noted in OECD [7], 'policymakers often have a linear view of the world. This approach ignores how systems interact and how their systemic properties shape interaction, leading to an over-emphasis on a limited set of characteristics'.

A systems thinking perspective on COVID-19 brings to the forefront the inherent interdependencies and interconnectivity that exists across national societal public health infrastructure. Jackson [5] defines the systems thinking paradigm as '…a discipline for seeing the 'structures' that underlie complex situations, and for discerning high from low leverage change…Ultimately, it simplifies life by helping us to see the deeper patterns lying beneath the events and the details'. Systems thinking also provides an approach to challenge our inherent beliefs and assumptions regarding causality and the impact of intervention strategies.

As described in Hynes et al. [4], 'systems thinking not only improves multi-disciplinary, cross-sectoral collaboration, it can also provide insights into systems behaviour and management by rigorous analysis of such aspects as system dynamics, feedback, sensitivity, and non-linear responses; the emergence of systems behaviour and properties; the optimization of system performance over different time horizons or for different groups; the anticipation and assessment of systemic risks; and the strengthening of resilience to external change and shocks'. Insights generated from the application of systems thinking and in particular System Dynamics make these approaches a key lens to support sensemaking.

Weick [8] refers to sensemaking in terms of '…how we structure the unknown so as to be able to act in it. Sensemaking involves coming up with a plausible

---

[1] https://health-infobase.canada.ca/covid-19/epidemiological-summary-covid-19-cases.html?stat=num&measure=total&map=pt#a2.

understanding—a map—of a shifting world; testing this map with others through data collection, action, and conversation; and then refining, or abandoning, the map depending on how credible it is' [1]. Ancona [1] describes sensemaking as an '… activity that enables us to turn the ongoing complexity of the world into a "situation that is comprehended explicitly in words and that serves as a springboard into action" (Weick et al. [9], p. 409). Thus sensemaking involves—and indeed requires—an articulation of the unknown'. Making sense of a 'complex' world and all that characterizes 'complexity' is a key theme in the sensemaking and crisis leadership literature.

System effects have been highlighted with regards to Covid-19 and '… how subjective or cultural factors such as trust in institutions and willingness to follow their advice and instructions, the sentiment of belonging to a community or the type of neighbourhood, can influence how a disaster unfolds' [7]. With this being said, systems thinking as a sensemaking lens reveals the key drivers, interactions, and dynamics of the societal systems. These are critical insights in which to model the efficacy of intervention strategies.

In this chapter, a case study will be presented of sensemaking about the COVID-19 pandemic in Canada in the Spring of 2020 using a System Dynamics model with the data available at that time and employing assumptions that needed to be made at the time about the effectiveness of future interventions.

2   Modelling

System Dynamics is a computer simulation technique that is based on solving a highly interconnected system of differential equations.[2] A model can be viewed as a series of Stocks (boxes) that contain some entities and Flows (pipes with valves) connected to these Stocks which increase and decrease the Stocks over time. Control variables use the information on the Stocks to manage the Flows. The System Dynamics model we used for COVID-19 in Canada was the simple SIR (susceptible-infected-recovered) and can be visualized using the special purpose software Vensim[3] (see Fig. 1). We also implemented the model in Microsoft Excel[4] to do the data fitting and present the results in this study.

The model begins with an initial number of infected people and an initial number of susceptible people. There is an uncontrolled reproduction rate for the virus that is affected by physical distancing and testing and isolating infected people to obtain the actual reproduction rate. This reproduction rate is divided by the illness duration to obtain a reproduction rate per day per infected person. This value is multiplied by the number of people infected and the fraction of people still susceptible to determine the outflow from the susceptible population and into the infected population. The

---

[2] https://www.systemdynamics.org/what-is-sd.

[3] https://vensim.com/coronavirus/.

[4] Albright, S. Christian, "VBA for Modelers: Developing Decision Support Systems with Microsoft Excel", 5th Edition, South-Western College Publications, 2015.

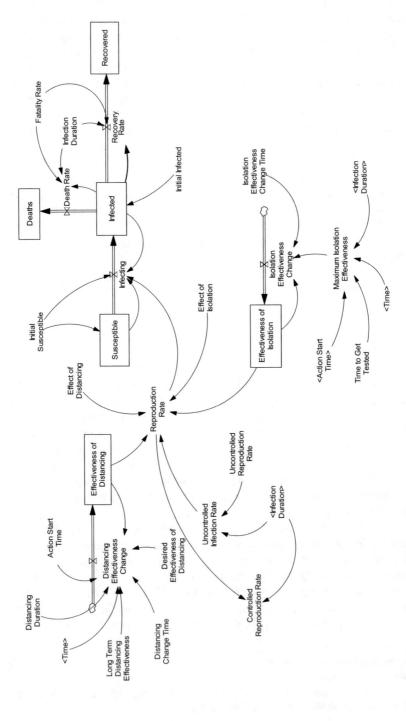

Fig. 1 Stock and flow diagram of a simple SIR epidemic model

infected people will eventually flow into either the deaths or the recovered groups based on the fatality rate divided by the duration of the illness.

We assume testing/isolation and physical distancing regulations come into effect on the same day but take different lengths of time to fully implement. The distancing regulation is for a limited amount of time. When the distancing regulation is removed, people may continue to voluntarily distance. On the other hand, the isolation regulation is assumed to be based on testing for the virus. The testing/isolation regulation will continue indefinitely. These are the short-term and long-term approaches to reducing the reproduction rate. The goal is to lower the reproduction rate to a value that is less than 1.0 and keep it there.

This SIR model is simple but a well-established epidemic model. We will use data from the Government of Canada[5] to estimate the parameters for the model. The equations used in the model are provided in Annex A.

3 Assumptions

We assume the duration of the illness averages 21 days. We assume the distancing takes effect on April 1st and lasts for 30 days at which time the distancing effectiveness falls back to 30%. We assume the goal is to isolate anyone who tested positive two days after they become infected. However, we assume it takes 60 days to gradually implement the testing required to allow for this policy to work. The goal of testing in two days was chosen to reduce the long term reproduction rate to a value less than 1.0.

4 Estimating the Model Parameters

Early in the pandemic we collected time series data for the Canadian provinces. We used the data up until March 31st to estimate the uncontrolled reproduction rate and the fatality rate. We found the parameters for the model by minimizing the squared error between the time series data on confirmed cases and deaths against the behaviour over time graphs from the model. We assumed that the distancing regulations were put in place on April 1st and used the data up until April 25th to estimate the effectiveness of the distancing at reducing the reproduction rate in the model.

We modelled each Canadian province separately. The graphs shown in Figs. 2 and 3 are examples of the fit of the model to the time series for Manitoba. It appears the distancing regulations were working well in Manitoba.

We summed up the results from each province to get an overall view of the Canadian situation. This neglected the Canadian Territories. However, the territories

---

[5]https://www.canada.ca/en/public-health/services/diseases/coronavirus-disease-covid-19.html.

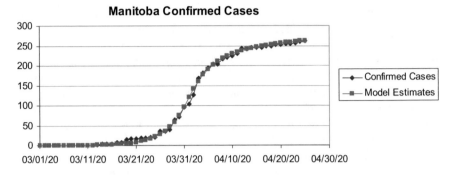

Fig. 2 Confirmed cases time series for manitoba and model behaviour over time

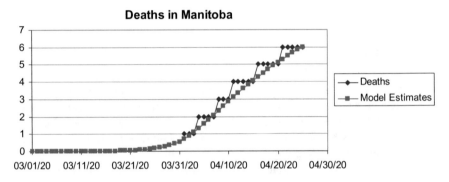

Fig. 3 Deaths time Series in manitoba and model behaviour over time graph

had very few cases and no death. In Figs. 4 and 5, we show the results of this summation.

It does not appear distancing is having much impact on Canada as a whole. This is because the Canadian situation is dominated by the two most populated provinces

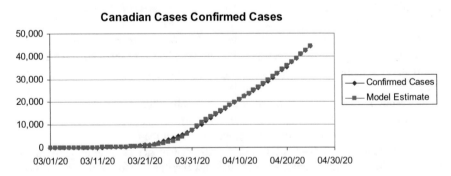

Fig. 4 Confirmed cases and model estimates

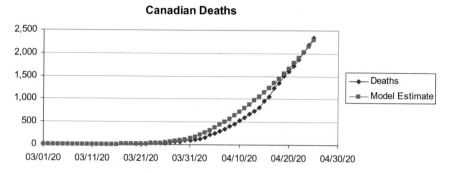

Fig. 5 Deaths and model estimates

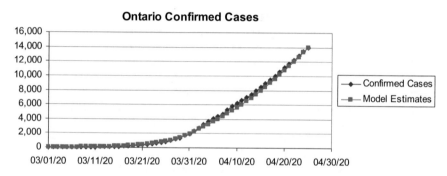

Fig. 6 Confirmed cases in Ontario and model estimates

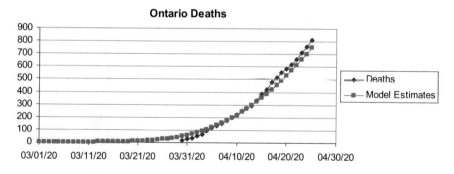

Fig. 7 Deaths in Ontario and model estimates

Ontario and Quebec. Figures 6 and 7 show the data fits for Ontario and Figs. 8 and 9 show the data fits for Quebec. The data fits for the other provinces is provided in Annex B.

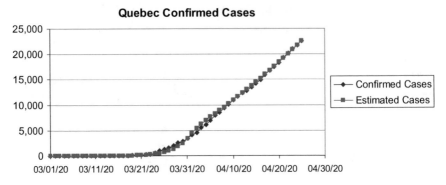

Fig. 8 Confirmed cases in Quebec and model estimates

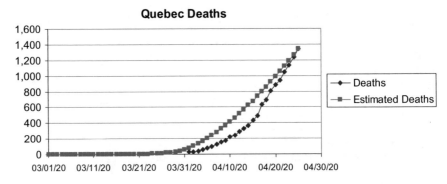

Fig. 9 Deaths in Quebec and model

Table 1 Model parameters for the Canadian Provinces

Province	Reproduction rate	Fatality rate	Distancing effectiveness (%)	Time to implement distancing
British Columbia	4.20	0.083	77	1
Alberta	6.93	0.040	67	1
Saskatchewan	6.95	0.018	90	1
Manitoba	6.21	0.036	94	3.76
Ontario	4.26	0.127	39	4
Quebec	7.90	0.129	73	1.80
New Brunswick	5.51	N/A	93	1.74
Prince Edward Island	4.78	N/A	92	1
Nova Scotia	6.07	0.043	60	1.67
Newfoundland and Labrador	7.50	0.018	94	1

Table 1 shows the parameters for the Canadian provinces.[6] There have been no deaths in New Brunswick or Prince Edward Island. Therefore, we could not calculate a fatality rate for those provinces.

5 Long Term Projections

System Dynamics attempts to model the physics of a problem and hypothesizes a cause and effect relationship. The approach also allows the possibility of a non-linear projection with which to evaluate policy options. We have seen the distancing regulations were effective in several Canadian provinces. Our baseline assumption was that the distancing regulations would be officially removed after 30 days (i.e. May 1st) while some voluntary distancing would remain at a lower effectiveness level (30%).

We also made assumptions about the potential of testing and isolating coming into effect slowly after April 1st and continuing indefinitely. The baseline assumptions were that once it was fully implemented people with COVID-19 would be tested on average two days after becoming infected and immediately isolated. We assumed it would take 60 days to fully implement this policy after the regulations came in place on April 1st (i.e. at the end of May).

Figures 10 and 11 are the projections for Canada under the baseline assumptions.

Recall in Fig. 2 Manitoba appeared to have the problem under control. One might ask what was projected to happen in Manitoba once the distancing regulation is lifted. Figures 12 and 13 show that the virus appears to be only temporarily under control during the distancing regulation period.

Table 2 shows the long term reproduction rate, the projected deaths, and cases in each province at the end of the year with the baseline assumptions.

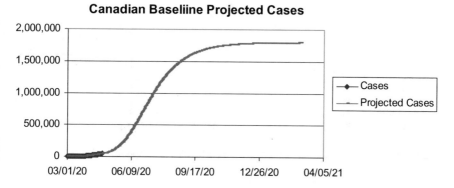

Fig. 10 Projections for cases under the baseline assumptions

---

[6]We assumed it took a minimum of 1 day to implement the distancing regulations.

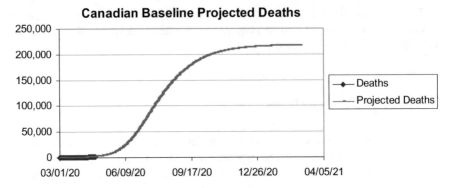

Fig. 11 Projections for deaths under the baseline assumptions

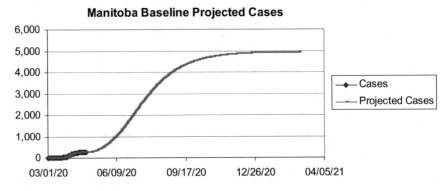

Fig. 12 Projections for cases in Manitoba under the baseline assumptions

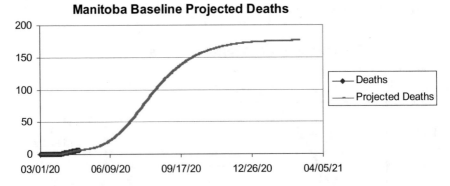

Fig. 13 Projections for deaths in Manitoba under the baseline assumptions

Table 2 Projections for Canadian Provinces under the baseline assumptions

Province	Long term reproduction rate	Projected deaths January 1st, 2021	Projected cases January 1st, 2021
British Columbia	0.49	692	8,350
Alberta	0.81	16,773	423,067
Saskatchewan	0.81	384	21,100
Manitoba	0.73	174	4,899
Ontario	0.50	15,417	121,849
Quebec	0.92	411,055	3,190,139
New Brunswick	0.64	0	1,214
Prince Edward Island	0.64	0	142
Nova Scotia	0.56	1,634	38,046
Newfoundland and Labrador	0.88	443	25,553

6 Testing the Assumptions

Recall we assumed, the regulations would stay in effect for 30 days, there would be a voluntary distancing after the regulations were lifted that was 30% effective, the goal would be testing 2 days after the people are infected and then isolating them but it would take 60 days to reach this testing goal. We will consider how important these assumptions are. We will change the values associated with each of these individual assumptions one at a time and hold the other parameters at their baseline values. In each case, we projected the deaths by January 1st, 2021.

First, we will look at the results when we vary the days of distancing. Recall that we assumed the regulated days of distancing was 30 days. From Fig. 14, we can see the model projects that hundreds of thousands of lives would be saved by distancing this length of time. If the distancing regulations could remain in place for 60 days at

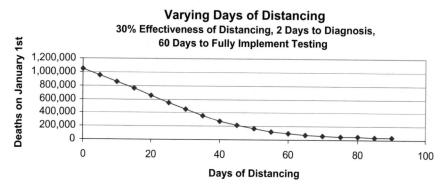

Fig. 14 Impact of varying regulated days of distancing

the current effectiveness levels, this would appear to be ideal. Longer than 60 days may not be necessary according to the projections of the model.

If Canada went completely back to normal with no distancing after the 30 days of the regulations, then approximately 1.1 million deaths were projected by the model. The baseline was that Canadians would voluntarily obtain distancing effectiveness of 30% but still, 450,000 deaths were projected by January 1st, 2021. If the voluntary distancing after the regulations come off could be maintained at 60% or more, the model projected very few deaths in the future (Fig. 15).

The model projected that the days to diagnosis has to be very low to have a significant impact on the deaths in the future. The baseline value was assumed to be 2 days fully implemented in 60 days. Anything larger than was projected to have catastrophic effects (see Fig. 16).

From Fig. 17, it would appear ideal if the testing and isolating were implemented in 30 days or less (i.e. on or before May 1st). Any value greater than 40 days (i.e. after mid-June) was projected to result in a significant increase in the number of deaths by January 1st, 2021.

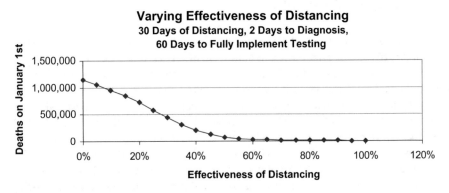

Fig. 15 Impact of varying long term effectiveness of distancing

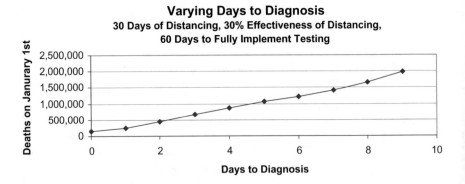

Fig. 16 Impact of varying days to diagnose

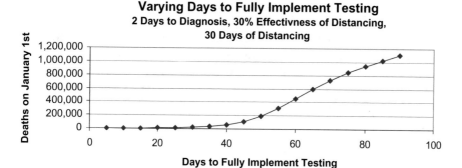

Fig. 17 Impact of varying days to fully implement testing

Next, we will consider the situation if we vary the assumptions at the same time (see Table 3).

We can see that the worst case is scenario 5 with a projected death of over 875,000 when the day of distancing is short, the voluntary distancing effectiveness is limited, the days to diagnosis is moderately high and the days to implement testing is high. The best-case scenario is number 12 with just over 7,000 deaths when the days

Table 3 Varying the assumed Parameters At The Same time

Scenario	Days of distancing	Voluntary distancing effectiveness (%)	Days to diagnosis	Days to implement testing	Deaths by January 1st, 2021
1	30	30	2	60	446,571
2	60	30	2	60	96,708
3	30	60	2	60	33,757
4	60	60	2	60	23,869
5	30	30	4	60	877,693
6	60	30	4	60	492,385
7	30	60	4	60	71,894
8	60	60	4	60	44,256
9	30	30	2	30	18,301
10	60	30	2	30	8,540
11	30	60	2	30	7,747
12	60	60	2	30	7,059
13	30	30	4	30	128,932
14	60	30	4	30	33,419
15	30	60	4	30	12,542
16	60	60	4	30	10,443

of distancing is high, the voluntary distancing effectiveness is high, the days to diagnosis is low and the days to fully implement testing is low. Scenarios 9 through 12 all have a relatively low number of projected deaths (i.e. under 20,000) which suggests reducing the days to diagnosis and days to fully implement testing are of primary importance. Comparable results (i.e. deaths under 15,000) can be achieved if the voluntary distancing effectiveness is quite high and the days to fully implement testing is low even when the days to diagnosis are not minimized.

7 Conclusions

The inherent interconnectivity and interdependency within societal public health systems require analysis that provides a deep understanding regarding the potential impact of COVID-19 on populations in response to intervention strategies. In dynamic systems, the effects of an intervention are only evident after a time delay. Understanding the system and its inherent dynamics is a key requirement for sense-making and is a game-changer in supporting crisis management of complex issues such as a global pandemic.

Early in the pandemic, we built a System Dynamics model of the standard SIR (susceptible-infected-recovered) approach to an epidemic. We modified the simple SIR model by including the impact of physical distancing, and testing and isolating infected people. We estimated the uncontrolled reproduction rate[7] and the fatality rate based on data available before March 31st, 2020. We assumed the physical distancing regulations were put in place on April 1st, 2020, and estimated the time to fully implement this distancing and the effectiveness of the distancing using data until April 27th, 2020. We found that that the results for Canada as a whole were dominated by the most populated provinces (Ontario and Quebec) and there the distancing regulations were having limited impact.

We made assumptions about how long the distancing would be in place, the effectiveness of voluntary distancing after the regulations are lifted, the intended number of days it would take to test and isolate infected people, and the time it would take to fully achieve this testing and isolating goal. We considered these values as a baseline for projecting the number of cases and deaths in the provinces and the country as a whole. We projected that the long term would see confirmed cases and deaths eventually leveling off around but at unacceptably high values.

At the time, a sensitivity analysis showed that being able to fully implement the testing and isolating of people soon after they become infected would have a significant impact on reducing the number of deaths to an acceptable level in the long term. If the days to diagnosis an infection could not be reduced as much as we were hoped but the testing could be fully implemented quickly, then improving the effectiveness of voluntary distancing after the regulations are lifted would be able to reduce the projected deaths to an acceptable level.

---

[7] Assuming the average duration of the illness is 21 days.

This case study demonstrates that caution needs to be used when projecting long term impact from small data sets using this technique. This is an example of the classic "butterfly" effect in this type of modelling, namely the sensitivity of the long term projections to inaccurate estimates of the initial conditions. This case study shows the value of testing the assumptions that are made in the long term projections. We feel that even with limited data and making highly sensitive assumptions, this approach appears valuable in sensemaking early in a health crisis like the COVID-19 pandemic in Canada.

ANNEX A: COVID-19 System Dynamics Model Equations

Action Start Time = 30
    Units: day
    Confirmed Cases = Deaths + Infected + Recovered
    Units: people
    Controlled Reproduction Rate = Reproduction Rate*Infection Duration
    Units: people/person
    Death Rate = Infected*Fatality Rate/Infection Duration
    Units: people/day
    Deaths = INTEG (Death Rate, 1)
    Units: people
    Desired Effectiveness of Distancing = 0.94
    Units: dmnl
    Distancing Change Time = 2
    Units: days
    Distancing Duration = 30
    Units: days
    Distancing Effectiveness Change = IF THEN ELSE(Time < Action Start Time,0, IF THEN ELSE(Time <=(Action Start Time + Distancing Duration),(Desired Effectiveness of Distancing-Effectiveness of Distancing)/Distancing Change Time, (Long Term Distancing Effectiveness-Effectiveness of Distancing)/Distancing Change Time))
    Units: dmnl/day
    Effect of Distancing([(0,0)-(1,1)], (0,1), (0.5,0.5), (1,0))
    Units: dmnl
    Effect of Isolation([(0,0)-(1,1)], (0,1), (0.5,0.5), (1,0))
    Units: dmnl
    Effectiveness of Distancing = INTEG (Distancing Effectiveness Change,0)
    Units: dmnl
    Effectiveness of Isolation = INTEG (Isolation Effectiveness Change,0)
    Units: dmnl
    Fatality Rate = 0.036
    Units: fraction

FINAL TIME = 365
Units: day
The final time for the simulation.
Infected = INTEG (Infecting-Death Rate-Recovery Rate, Initial Infected)
Units: people
Infecting = (Susceptible/Initial Susceptible)*Infected*Reproduction Rate
Units: people/day
Infection Duration = 21
Units: day
Initial Infected = 1
Units: people
Initial Susceptible = 1360396
Units: people
INITIAL TIME = 0
Units: day
The initial time for the simulation.
Isolation   Effectiveness   Change   =   (Maximum   Isolation   Effectiveness-
Effectiveness of Isolation)/Isolation Effectiveness Change Time
Units: dmnl/day
Isolation Effectiveness Change Time = 60
Units: days
Long Term Distancing Effectiveness = 0.3
Units: dmnl
Maximum Isolation Effectiveness = IF THEN ELSE(Time <=Action Start Time,
0, (Infection Duration-Time to Get Tested)/Infection Duration)
Units: dmnl
Recovered = INTEG (Recovery Rate, 0)
Units: people
Recovery Rate = Infected*(1-Fatality Rate)/Infection Duration
Units: people/day
Reproduction   Rate   =   Uncontrolled   Infection   Rate*Effect   of
Distancing(Effectiveness   of   Distancing)*Effect   of   Isolation(Effectiveness   of
Isolation)
Units: people/person/day
SAVEPER = TIME STEP
Units: day [0,?]
The frequency with which output is stored.
Susceptible = INTEG (-Infecting, Initial Susceptible)
Units: people
TIME STEP = 0.125
Units: day [0,?]
The time step for the simulation.
Time to Get Tested = 2
Units: days

Uncontrolled Infection Rate = Uncontrolled Reproduction Rate/Infection Duration

Units: people/person/day

Uncontrolled Reproduction Rate = 6.41

Units: dmnl

ANNEX B: Fitting the Model to Canadian Provinces

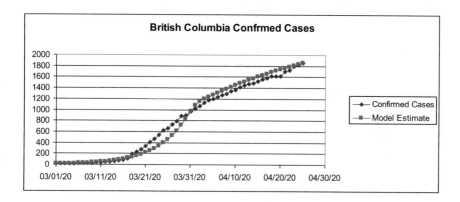

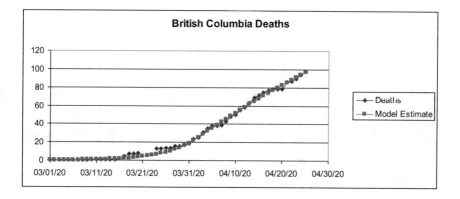

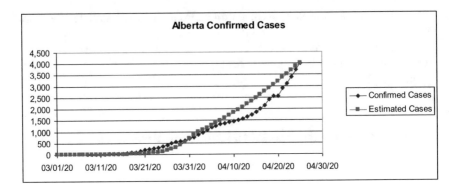

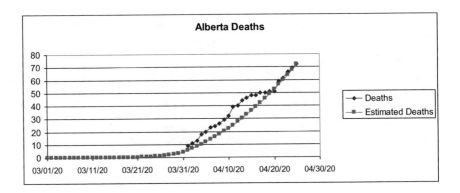

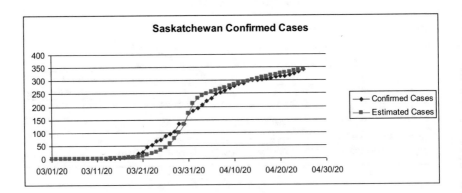

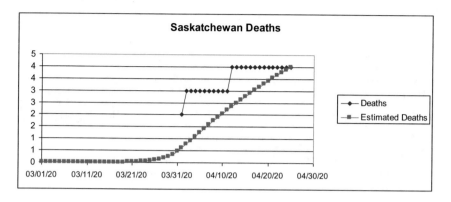

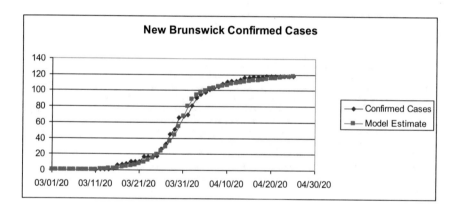

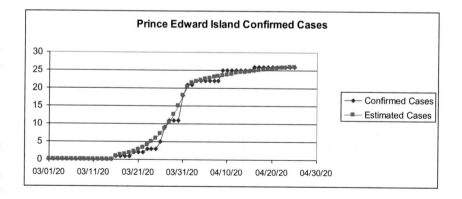

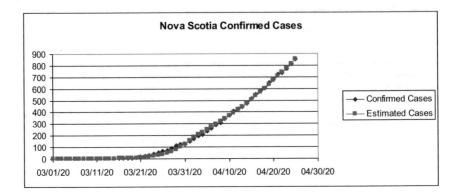

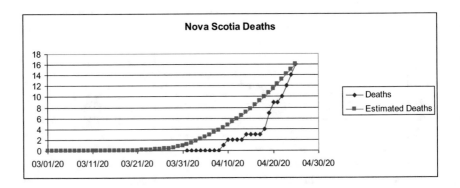

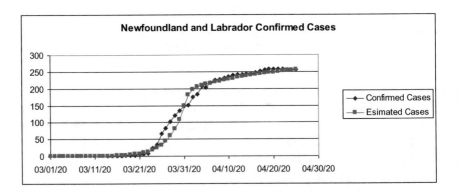

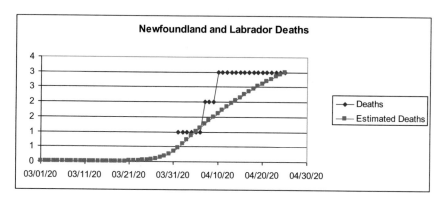

References

1. Ancona DL (2011) SENSEMAKING framing and acting in the unknown in handbook of teaching leadership, 3–20. Sage Publications Inc, Thousand Oaks, CA
2. Helbing D (2012) Systemic risks in society and economics. In: Social self-organization. Springer, Berlin, Heidelberg, pp 261–284
3. Heyman et al (2015) Global health security: the wider lessons from the west African Ebola virus disease epidemic, vol 385, May 9, 2015 www.thelancet.com
4. Hynes W, Lees M, Muller JM (2020) Systemic thinking for policy making: the potential of systems analysis for addressing global policy challenges in the 21st century. https://www.oecd-ilibrary.org/governance/systemic-thinking-for-policy-making_879c4f7a-en
5. Jackson MC (2003) Systems thinking: creative holism for managers. Wiley, West Sussex
6. Linkov I, Trump BD, Hynes WO (2019) Resilience-based strategies and policies to address systemic risks. SG/NAEC(2019)5. https://www.oecd.org/naec/averting-systemic-collapse/SG-NAEC(2019)5_Resilience_strategies.pdf
7. OECD (2020) A systemic resilience approach to dealing with Covid-19 and future shocks new approaches to economic challenges (NAEC), 28 April 2020
8. Weick KE (1995) Sensemaking in organizations. Sage, Thousand Oaks, CA
9. Weick KE, Sutcliffe KM, Obstfeld D (2005) Organizing and the process of sensemaking and organizing. Organ Sci 16(4):409–421
10. Weick KE, Sutcliffe KM (2007) Managing the unexpected: resilient performance in an age of uncertainty, 2nd edn. Wiley, San Francisco

How Politics Shapes Pandemics

Simon Bennett

Abstract Academics claim that past events hold important lessons for risk managers. This chapter is based on that premise, specifically on the possibility that those charged with meeting the challenge of the COVID-19 pandemic can learn lessons from the successes and failures of those who met the challenge of the 1918–1919 Spanish flu pandemic. One of the most interesting aspects of the United States of America's (USA's) experience of the Spanish flu pandemic was the degree to which political agendas shaped local responses. Some cities prioritised public health over almost every other consideration. Others prioritised the local economy and the war effort. Cities that prioritised public health generally experienced less death and suffering than those that did not. Early Twentieth Century politics influenced how the Spanish flu pandemic proceeded. Today, New Millennium politics is shaping the COVID-19 pandemic. Plus-ça-change. During the early stages of the pandemic, some countries prioritised public health. Others did not. Regarding numbers of excess deaths attributable to COVID-19, there are sharp differences between the best and worst performing countries. Academics have long argued that Modernity is Janus-faced, gifting us technologies that, contingent on how they are used, either enhance or erode public safety. Consider, for example, nuclear fission and dynamite. It is a sad irony that one of Modernity's greatest achievements—affordable air service—helped seed COVID-19 across the globe. The COVID-19 pandemic is an echo from the past. Dark Age dystopias—and mentalities—persist. The promissory note that is Modernity is being questioned as never before.

Keywords Spanish flu pandemic · Passive learning · Active learning · Politics · Priorities · COVID-19

S. Bennett (✉)
Civil Safety and Security Unit (CSSU), University of Leicester, Leicester, UK
e-mail: sab22@leicester.ac.uk

© The Author(s), under exclusive license to Springer Nature Switzerland AG 2021
A. J. Masys (ed.), *Sensemaking for Security*, Advanced Sciences and Technologies
for Security Applications, https://doi.org/10.1007/978-3-030-71998-2_12

1 Introduction

One of the most powerful and persistent narratives of our age is that Modernity—a key aspect of which is science-informed social, economic and political development—has transformed the world in which we live from a place of superstition, danger and fear to a place of rationality, safety and contentment [1]. As a variety of recent events demonstrate, this narrative is a conceit that hides a more complex and, from a security standpoint, less satisfactory picture. Modernity's successes should be celebrated. Its failures closely studied.

1.1 Modernity and Freedoms

At the time of writing, Afghanistan's fragile social and political institutions, shaped by Modernist values such as universal suffrage, democratic accountability, equality of access to education and freedom of belief and expression, are under attack from the Islamist Taliban, a terrorist faction determined to create a system of governance grounded in superstition, irrationality, prejudice, gender discrimination, sexism and other Dark Age habits. Nothing demonstrates the barbarism of the Taliban's perverted credo better than the attempted assassination, in 2012, of Malala Yousafzai, a fifteen-year-old Pakistani schoolgirl who had publicly supported female access to education. Malala was shot in the head in front of her school friends by a Taliban gunman. In 2008, the Taliban, having colonised Pakistan's Swat Valley, called for the ending of female access to education. The terrorists then proceeded to blow up many of the Swat Valley's schools. Barbarous acts committed by the Taliban during their reign of terror included the execution of a dancer accused of immorality (her body was put on public display to terrorise) and the flogging of a seventeen-year-old girl accused of having 'illicit relations' with a man [2–4]. During the Cold War, the mujahideen—the precursor to the Taliban—were supplied with arms by the Central Intelligence Agency (CIA) to improve their chances of defeating the Red Army. A case of 'my enemy's enemy is my friend?' No episode better demonstrates the amorality of super-power politics [5, 6]. Has the threat posed by religious fundamentalism to secular democracy and associated institutions receded? Absolutely not. In the English city of Birmingham, alleged attempts by Islamists to infiltrate and coerce state-funded, secular schools have been investigated. One report, produced by the former head of the Metropolitan Police Service's (MPS's) counterterrorism command, observed: "I found clear evidence that there are a number of people, associated with each other and in positions of influence in schools and governing bodies, who espouse, endorse or fail to challenge extremist views .... [S]ome governors ... used the argument about raising standards to justify increasing the influence of faith in ... schools" [7].

1.2 Modernity and Health

Modernity, underpinned by the scientific method, has, thanks to medical science and service-delivery and public health vectors such as Britain's National Health Service (NHS), the United States of America's Centres for Disease Control and Prevention (CDC), the European Centre for Disease Prevention and Control (ECDC) and the World Health Organisation (WHO), produced startling successes in the field of public health. Consider, for example, the fate of that most cruel of diseases, polio. It was polio that struck down one of America's most able and successful politicians, US President Franklin D. Roosevelt (Fig. 1) and one of Britain's most talented rock musicians, Ian Dury. Polio can be lethal.

Most often, the virus, contracted from infected faeces or aerosol droplets launched in coughs and sneezes, leaves the sufferer with "muscle weakness, shrinking of the muscles (atrophy), tight joints (contractures) [or] deformities, such as twisted feet or legs" [9]. The 1950s saw two major breakthroughs, with, in 1953, the development by Jonas Salk of the formalin-inactivated vaccine (IPV) and the development by Albert Sabin in 1956 of live-attenuated vaccines (OPV). In 1988, the WHO proposed ridding the world of polio. By April 2017, the public–private Global Polio Eradication Initiative (GPEI) had reduced polio infections by 99% [10, 11]. Underwritten by medical science and propelled by political will, organisational skills and hard cash, the GPEI has triumphed. A victory for Modernity.

The past is no guide to the future, however [12]. In recent times, the drive to eradicate disease has suffered several reverses, with pernicious viruses such as Ebola Virus Disease (EVD) and Severe Acute Respiratory Syndrome (SARS) causing suffering

Fig. 1 President of the United States and polio victim, Franklin Delano Roosevelt, photographed in August 1944. Exhausted by the war, the greatest modern American died in April 1945. [8]

and death in affected populations. Ebola Virus Disease, discovered in 1976, originates in animals and is often fatal in humans: "The virus is transmitted to people from wild animals and spreads in the human population through human-to-human transmission. The average EVD case-fatality rate is around 50%. Case-fatality rates have varied from 25 to 90% in past outbreaks". West Africa saw a severe outbreak in 2014–2016. The virus surfaced in the Democratic Republic of Congo between 2018 and 2019 [13]. The SARS virus, which originated in China, caused two pandemics between 2002 and 2004. While not as lethal as EVD, SARS caused death, discomfort and disruption: "During the period of infection, there were 8,098 reported cases of SARS and 774 deaths. This means the virus killed about 1 in 10 people who were infected. People over the age of 65 were particularly at risk, with over half of those who died from the infection being in this age group …. The SARS pandemic was eventually brought under control … following a policy of isolating people suspected of having the condition and screening all passengers travelling by air from affected countries for signs of the infection" [14].

The most dramatic reversal to date has been caused by the 2019-ongoing COVID-19 pandemic. The Chinese claim they first detected the virus in the city of Wuhan in late 2019. Thanks to the efficiency of air transport, and proclivity of people to explore, the virus spread around the globe. On 11 March, 2020, the WHO declared a global pandemic. A pandemic sees a disease, such as COVID-19, passing from person to person on every continent. Following advice from the WHO and other agencies, governments implemented quarantines, lock-downs and social-distancing measures. On 20 May, 2020, the British Broadcasting Corporation's (BBC's) Visual and Data Journalism Team reported that: "Globally, more than 4.5 billion people … have been living under social distancing measures …. Those restrictions have had a big impact on the world economy …. The United Nations World Food Programme has … warned that the pandemic could almost double the number of people suffering acute hunger" [15]. On 14 April, 2020, the International Monetary Fund's (IMF's) Economic Counsellor and Director of Research observed: "The magnitude and speed of collapse in [economic] activity that has followed [the lockdowns] is unlike anything experienced in our lifetimes …. [M]any countries now face multiple crises—a health crisis, a financial crisis and a collapse in commodity prices, which interact in complex ways …. This makes the Great Lockdown the worst recession since the Great Depression, and far worse than the Global Financial Crisis" [16]. The virus hit Britain hard, causing significant numbers of excess deaths (due in part to the Johnson government's abject failure to protect the frail elderly in care homes) and severe economic dislocation. Towards the end of May 2020, the Chancellor of the Exchequer warned that the United Kingdom was facing a "severe recession, the likes of which we haven't seen" (Sunak cited in [17]). Prior to Sunak's gloomy prognosis, successful British companies had been laying off workers. The pandemic-induced lock-down hit Britain's aviation sector hard, with airlines such as British Airways, Virgin Atlantic and easyJet, and suppliers such as Rolls Royce, laying off significant numbers of workers [18].

1.2.1   Understanding Public Health Failures

The reasons for such reversals are complex. Viewed through a systems-thinking lens [19], they originate in biological mechanisms, which are outside our control (we cannot prevent viruses carried by fauna from mutating into ever-more lethal strains) and in political mechanisms, which are, to a degree, controllable. While we can do little to reduce the prevalence of viruses such as EVD, SARS and COVID-19 in wildlife, we can—and should—do something about how outbreaks are perceived and managed by those we trust to keep us safe.

This analysis will test three hypotheses related to public health:

- 'That central and local governments, through timely and considered actions that prioritise public health, can reduce the death and suffering caused by pandemic'
- 'That important lessons can be learned about reducing death and suffering by studying how previous pandemics have been managed, for example, the three-wave Spanish Flu pandemic of September 1918 to February 1919'
- 'That Modernity will deliver ever more comfort and security'.

To give the analysis historical depth, the second hypothesis will be tested first.

2   Testing the Three Public Health Hypotheses

2.1   *Active Learning as a Mitigative*

Toft [12, 20] argues that the way in which threats have been managed in the past can help those responsible for public safety (legislators, civil servants, emergency services personnel, voluntary workers, etc.) to better manage current threats—even when those threats are different in nature and scale. Toft distinguishes two types of learning: passive and active. In the former, lessons are drawn but not actioned. In the latter, lessons are actioned. Bennett [21, 22] argues that interceding social, economic and political factors may inhibit active learning. "[T]he timely application of preventative or mitigative measures may be inhibited by the intercession of such social, economic and political factors as waning public and/or political interest, financial constraints or changes in the political agenda" claims Bennett [22, p. 3]. Buchanan [23, p. 273] observes: "[F]ollowing an accident, crisis, disaster or other extreme event, the recommendations from investigations and inquiries are often not implemented. The failure to change can then lead to a repeat of that event". However obvious and urgent the lessons, active learning cannot be guaranteed.

2.1.1 Opportunities for Active Learning

The Spanish flu pandemic of 1918–1919 infected around one third of the world's population and killed an estimated twenty to fifty million people (some estimates put the death toll at one hundred million) [24, 25]. No health event before or since has sowed so much misery and destruction: "[T]he pandemic killed more people in a year than AIDS has killed in 40 years, more than the bubonic plague killed in a century" [25].

Pandemics happen when a novel virus is introduced into a population that has no immunity to it. The influenza virus, which is easily transmitted, can mutate rapidly, "changing enough that the human immune system has difficulty recognising and attacking it even from one season to the next" [25]. Vaccines began to be developed in the early Twentieth Century [26]. However, there were no antivirals or vaccines for the Spanish flu.

Those most at risk from the flu virus include "Young children, people over age 65, pregnant women and people with certain medical conditions, such as asthma, diabetes or heart disease". The health impacts on the most vulnerable include "severe pneumonia, ear and sinus infections and bronchitis" [27].

In the federal United States, where Spanish flu eventually killed around 670,000 [25], health-protection strategies varied from city to city (Fig. 2). Some city authorities introduced strict control measures. Others took a more relaxed approach. The residents of these cities paid a heavy price for civic leaders' complacency.

In the United States, the Spanish flu appeared in three waves. The first, in the spring of 1918, was relatively mild (Fig. 3). Despite the virus's novelty, there were few recorded deaths. The majority of those infected recovered within days [25]. Civic authorities should have read the first wave as a warning. The second wave, which appeared in the autumn of 1918, was more deadly: "Victims died within hours or days of developing symptoms, their skin turning blue and their lungs filling with

Fig. 2 Seattle police officers photographed in December 1918. The officers wore masks supplied by the red cross. Cities responded to the pandemic in different ways. [28]

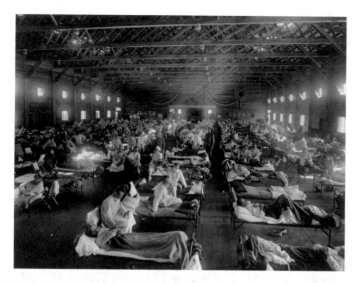

Fig. 3 Camp Funston, at Fort Riley, Kansas, in 1918. The Spanish flu hit the military hard, with more American soldiers dying from the virus than on the battlefields of Europe. [30]

fluid that caused them to suffocate" [27]. It took until 2008 for scientists to discover what made the virus so deadly: "A group of three genes enabled the virus to weaken a victim's bronchial tubes and lungs and clear the way for bacterial pneumonia" [27]. At the time of the pandemic, doctors knew next to nothing about its causes, and even less about how to help sufferers. Quack treatments and remedies abounded [29].

The pandemic could not have occurred at a worse time. The United States had entered the Great War—the war to end all wars—on the side of the Allies on April 6, 1917. The pandemic commenced in early 1918. With the country at war, politicians' and bureaucrats' attention was focused on supporting the war effort. Money had to be raised. The production of war materiel was prioritised. Doctors and nurses—a vital resource in any pandemic—were recruited to the military: "The war had drained doctors and nurses away to the battlefields, leaving many communities dangerously shorthanded" [29, p. 224].

2.1.2 Philadelphia's Response to the Spanish Flu

Philadelphia's response to the threat to public health posed by the Spanish flu was inadequate. Despite its proximity to Boston, which was already suffering a major outbreak, Philadelphia's political class and sinecure bureaucrats took few precautions, save issuing reassurances. Philadelphia's Director of Public Health and Charities, Dr. Wilmer Krusen, was a gynecologist who knew little about the science of epidemiology. Krusen issued reassurances grounded not in hard data but in a hopelessly misplaced optimism that everything would work out for the best. He insisted the authorities would be able to "confine this disease to its present limits" and professed "no concern". When the pandemic hit hard, Krusen simply lied, claiming that Philadelphians were being struck down by "old-fashioned influenza or grippe". As Philadelphia's meagre public health infrastructure struggled to cope, Krusen claimed that "the disease will decrease" [31]. Krusen's chronic denialism lulled the city authorities into a false sense of security. Despite mounting evidence that the Spanish flu could wreak as much havoc in Philadelphia as it had in Boston, the authorities sanctioned a parade to promote the Fourth Liberty Loan Drive. The parade, held on September 18, 1918, saw thousands of patriotic Philadelphians line the streets in support of America's war effort. The public health consequences were dire: "Within 10 days, over 1,000 Philadelphians lay dead, with another 200,000 estimated ill …. By March 1919, when the threat of influenza lifted, Philadelphia had lost over 15,000 of its citizens" [31]. While Boston's worst daily death toll stood at 202, Philadelphia's stood at 759 [25, 31].

Fairness requires that Philadelphia's lamentable response to the health emergency be evaluated in a wider context. Viewed through a systems-thinking lens [19], it is likely that the decisions of civic leaders and other key players were influenced by several micro, meso and macro social, economic and political factors. For example:

- The federal government's strongly expressed desire that the American people support the war effort and fund-raising mechanisms such as Liberty Loans. The President of the United States, Woodrow Wilson, had invested significant political capital in building support for the war effort. He remained silent on the pandemic [31]. Frequently, those in positions of authority nationwide—Governors, Mayors, public health officials—set entirely the wrong example. At every level of government, the response to the pandemic was uneven [31]
- Patriotic Americans' desire to support troops fighting thousands of miles from home in the most appalling conditions
- The fact that Philadelphia, like many East Coast cities, hosted military personnel. Philadelphia's economy was buoyed by military expenditure
- The fact that America's entry into the Great War had seen an influx of workers into major conurbations such as Philadelphia. Population density increased: "The city, whose infrastructure had buckled even before the war, was now teeming with new workers and migrants who stretched the housing stock, streets and transportation [system] to capacity" [31]
- Determined opposition from the business community to the shuttering of stores, saloons, theatres, restaurants and other outlets
- The fact that during the country's involvement in the Great War the federal authorities, through the Sedition Act, effectively censored the media, thereby stymying critique and debate. In Philadelphia, newspaper editors refused to print readers' letters and reporters' stories that questioned politicians' decisions [25]. The Fourth Estate is a guardian of justice, freedom and security. In countries where newspapers, the Internet and other media are subject to state controls or outright censorship (China, for example) the population is potentially less safe and secure [19, 32, 33].

However, despite these mitigations, the fact remains that Philadelphians paid a heavy price for civic leaders' patriotism and pro-business stance, and Dr. Wilmer Krusen's denialism, mendacity and ignorance of epidemiology.

2.1.3 Pittsburgh's Response to the Spanish Flu

At the time the United States entered the Great War, Pittsburgh was one of the country's most important centres of manufacture [34]. Steel City (Fig. 4), as Pittsburgh was known, was a cornerstone of the Allied war effort: "Industrious Pittsburgh, then the country's eighth-largest city and home to some of the largest manufacturers in the world, built the tools featured in the world's deadliest conflict …. Pittsburgh's war effort began on Dec. 30, 1914, when the British government approached Westinghouse Electric and Manufacturing Co. with an order for 3 million artillery shells. Westinghouse filled the order in 45 days, despite having no experience in manufacturing weaponry" [35]. Steel City's industrial base comprised over 250 places of manufacture, the city and its environs producing around half the munitions and steel used by the Allies (Britain, France, Russia and Serbia) during the Great War [34].

Opinion is divided as to how well Pittsburgh's civic leaders and bureaucrats defended the city's population against the Spanish flu. According to O'Toole [37], Pittsburgh's response was dulled by "petty self-interest and minimal coordination". According to the University of Michigan Centre for the History of Medicine

Fig. 4 At the time of the Great War and Spanish flu pandemic, Pittsburgh was the USA's Steel City, a significant producer of war materiel. The photograph, taken in 1906, shows Steel City's Strip District neighbourhood looking northwest from the roof of Union Station. Note the pollution [36]

Fig. 5 A shameful statistic.
Pittsburgh's death rate [38]

US city	Death rate per 100,000 residents after 24 weeks of pandemic
Boston	710
Philadelphia	748
Pittsburgh	807
The average death rate for Eastern cities was 555 per 100,000	

[38], Pittsburgh's response benefitted from "organised local leadership and efficient allocation of resources".

The truth of the matter may lie somewhere between these two readings. While it is true that civic leaders implemented protective measures [29], it is also true that these measures were implemented somewhat chaotically. For example, to deliver maximum protection, social distancing measures must be implemented simultaneously. Pittsburgh's civic leaders failed to do this: "Modern epidemiological studies of the 1918 epidemic [sic] show that the early, sustained and layered implementation of ... social distancing measures had an impact on the severity of a city's peak and overall influenza and pneumonia mortality. That is, it is not enough to simply utilise multiple measures; those measures should be implemented at the same time so as to create the maximum social distancing possible in a community" [38].

Whatever one's reading of Pittsburgh's response to the Spanish flu, the fact remains that Steel City suffered the highest death rate from the Spanish flu of any US city (Fig. 5). Indeed, Pittsburgh was the worst-hit city in the western world.

As per the evaluation of Philadelphia's response, fairness requires that Pittsburgh's response to the health emergency be evaluated in a wider context. Viewed through a systems-thinking lens [19], it is likely that the decisions of civic leaders and other key players were influenced by several micro, meso and macro social, economic and political factors. For example:

– Like Philadelphia's civic leaders, those in Pittsburgh would have been touched by the patriotic fervour fomented by Woodrow Wilson. Pittsburgh's role as a centre of manufacture would have made it a key source of war materiel
– Like its civic leaders, Steel City's business leaders would have been touched by President Wilson's call to arms, not least because wars are good for business [39, 40]
– The Sedition Act would have exerted a chilling effect on critique and debate. In time of war, it is natural for countries to set agendas and introduce measures that are believed to improve the chances of victory. For example, during the Second World War, the British government used the newly-created Ministry of Information to ensure the mass media did not broadcast or publish information useful to the enemy [41]

- Pittsburgh derived income and prestige from its heavy industries [34]. Civic leaders' determination to grow the city's industrial base meant that for many years public health had been sacrificed on the altar of economic success. According to O'Toole [37], there existed amongst the political class "a longstanding apathy toward the health of the general population". This apathy had serious consequences for public health: "Between 1900 and 1902, Pittsburgh recorded 253 deaths per 100,000 from pneumonia, behind only New York, while the years 1912 through 1914 saw the death rate for pneumonia rise to 261 per 100,000, the worst rate in the nation" (Higgins cited in [37]). Thus, it can be demonstrated that Pittsburgh's prosperity came at a price—the health of its inhabitants. Pittsburgh's poor air quality represented a latent error or resident pathogen (see Reason [42, 43] for a definition)—a circumstance capable of amplifying the harms inherent in disease, including influenza. Higgins (cited in [37]) observes: "Pittsburgh's horrendous air quality, the result of coke production and the burning of bituminous coal, was an important environmental factor that contributed to the severity of the illness in Pittsburgh's residents". Pittsburgh's poor-air-quality latent error derived largely from the self-interested behaviour of its industrialists: "The business community thwarted attempts to pass or enforce smoke abatement in Pittsburgh because of the added expense such measures entailed and because, philosophically, they resented governmental encroachment on private-property prerogatives" (Higgins cited in [37])
- Pittsburgh's decision makers operated within a system of government that evidenced a pro-industry bias. Inevitably, this bias manifested in appointment decisions, bureaucratic structures and decision-making. The practice of granting citizens membership of two or more boards introduced the possibility of conflicts of interest. The fact that the director of Pittsburgh's Board of Health, William H. Davis, was also a member of the Pittsburgh Chamber of Commerce might have influenced his decision-making vis-à-vis the city's response to the pandemic. Higgins (cited in [37]) describes Pittsburgh's Board of Health—which did not produce a single annual report between 1912 and 1919—as "a hollow municipal organ composed of political appointees".

While, at the time of the 1918–1919 pandemic, the behaviour of Pittsburgh's civic leaders may have been less reckless than that of Philadelphia's civic leaders, in the years leading up to the event, it left much to be desired. O'Toole [37] speaks of "decades of corrupt and short-sighted public policy". Certainly, in their determination to attract and retain prestigious manufacturing industries, the city fathers and their agents had so compromised public health that when the Spanish flu appeared in the city in October, 1918, the mass of the population was less resistant than it could and should have been. Underlying health problems amplify and accelerate the impacts of novel viruses [27]. Like any latent error, an underlying health problem, such as an above-average incidence of heart or respiratory disease, creates a vector for new infections. The University of Michigan Centre for the History of Medicine [38] notes: "Combined with the delay in closing schools, Pittsburgh's pollution may have contributed to the severity of its bout with influenza in 1918".

2.1.4 St. Louis's Response to the Spanish Flu

Like Philadelphia and Pittsburgh, St. Louis was vital to the war effort. The city's factories supplied materiel to the military, and its neighbourhoods men and women to fight and serve in France [44, 45]. Practical support for the war effort is the back-story to St. Louis's response to the Spanish flu pandemic of 1918–1919.

According to the University of Michigan Centre for the History of Medicine [46], thanks largely to the efforts of its Health Commissioner, Dr. Max C. Starkloff, St. Louis waged a vigorous war against the Spanish flu that saved lives and eased suffering. Starkloff was an immense figure in this campaign. Battered by vociferous business interests, church leaders, trades unions, transport companies, the Mayor's Office and the Federal government, Starkloff, supported by his loyal Assistant Health Commissioner, Dr. G.A. Jordan, and a knowledgeable, civic-minded and strong-willed medical advisory board, refused to yield: "[O]n November 9 [1918], [Starkloff] issued a sweeping proclamation immediately closing all non-essential stores, businesses and factories for four days in a drastic attempt to stamp out the epidemic once and for all .... The Retailers Association and the Chamber of Commerce complained loudly, and even the Federal government, concerned about war production, bore down on Starkloff .... In spite of the furore, people and businesses generally complied [with Starkloff's edict], though the beleaguered health commissioner did amend the proclamation to allow work on war contracts to continue" [46]. Starkloff's prioritisation of public health over everything bar war-work paid dividends (Fig. 6).

Some of the factors that contributed to St. Louis's excellent public health performance during the pandemic included:

– Robust, civic-minded leadership from the city's Health Commissioner
– The creation, in the form of the medical advisory board, of a credible source of independent advice on public health matters
– The Health Commissioner's collegiate management-style, that saw him canvass opinion on important decisions. Despite his determination to protect the public, Starkloff allowed those with vested interests to air their views in meetings. Starkloff's approach evidenced mindfulness and a penchant for high-reliability organising (see [47, 48] for definitions)
– Able support from loyal civil servants such as the Assistant Health Commissioner

Fig. 6 A heartening statistic. St. Louis's death rate during the Spanish flu [46]

City	Death rate per 100,000 residents
St. Louis	358
The average death rate for Eastern cities was 555 per 100,000	

– The creation by the Health Commissioner of a Bureau of Information to explain to residents what had been decided, and why. Starkloff understood the value of accurate and timely messaging during a crisis.

2.1.5 Lessons

Several lessons can be drawn from the responses to the Spanish flu pandemic described above:

– Civil servants responsible for public health who are, by virtue of multiple committee memberships, conflicted, may not do as much to promote public health as they ought
– Civil servants responsible for public health who, by virtue of their ignorance of epidemiology, lack the knowledge to prosecute an effective campaign against pandemic, are unfit for office and should be substituted (to borrow a popular phrase, they are unfit for purpose)
– The national interest may be defined in multiple ways. For example: It may be defined as protecting the public from a killer disease; It may be defined as promoting wealth generation; Or it may be defined as defeating an enemy on the battlefield. The case studies suggest that in time of war, public health is but one concern among many. More pressing concerns include: The timely supply of war materiel; The provision of doctors and nurses to provide care at the front for wounded soldiers
– The prioritisation of wealth generation over public health may so damage the latter that citizens have little or no resistance to novel viruses (such as the Spanish flu, EVD, SARS or COVID-19). In Pittsburgh, civic leaders' prioritisation of wealth generation helped create a city with high levels of atmospheric pollution. By the time the pandemic hit, Pittsburgh's pro-industry civic leaders had effectively normalised environmental degradation. Pittsburgh's health system was inadequate, a victim of chronic under-investment: "Pittsburgh had no health infrastructure of which to speak: just 20 community hospitals, some staffed by only one nurse, with an acute shortage of beds" [31]
– A government facing multiple threats may, in time of war, define the national interest solely in terms of security. Under such circumstances, other threats may not receive adequate attention. Further, the government in question may seek to convert those who define the national interest differently to its preferred world view. Pressure may be applied. If that fails, coercion may be used. It appears that little or no pressure was required to persuade Philadelphia's civic leaders to proceed with their Liberty Loan Drive parade—a gathering that accelerated the spread of the Spanish flu through the city
– Resisting pressure from vested interests requires strength, determination and a capacity for reasoned, objective decision-making. In St. Louis, Health Commissioner Starkloff, while open to suggestions from diverse interests, never lost sight of his primary duty—safeguarding. By contrast, in Pittsburgh, William H. Davis,

director of the Board of Health, was conflicted by virtue of his membership of the Pittsburgh Chamber of Commerce

- In a pandemic, it is vital that the public be provided with clear, unbiased and timely advice. In St. Louis, Commissioner Starkloff created the Bureau of Information, a mechanism he used to influence opinion and behaviour. At the outbreak of the Second World War, the British government created the Ministry of Information [41] for the same reasons—to influence opinion and behaviour
- Living conditions influence the dynamics of pandemic. In Philadelphia, an influx of workers had stressed the city's infrastructure. Housing and public transport were overstretched. The same conditions obtained in Pittsburgh: "The housing stock was … poor, with most working-class residents crammed into tenements that, under the right circumstances, were petri dishes for disease. They were better off than the 50,000 residents who lived in rooming houses, where they not only shared rooms with perfect strangers but also beds, which they occupied in shifts" [31]. Overcrowding facilitates the transmission of disease. Plus -ça -change. Today, the US aviation industry's crash-pad culture, which sees cash-strapped pilots, cabin crew and engineers renting a bed space by the week or day in severely overcrowded flats and houses in airport catchments [49], provides a vector for the transmission of disease.

2.2 In Time of Pandemic, Central and Local Governments that Prioritise Public Health Can Reduce Death and Suffering

While it is axiomatic that central and local governments, through timely and considered actions that prioritise public health, can reduce the death and suffering caused by pandemic, it is apparent from the case studies that central and local governments weigh public health against other considerations, such as economic prosperity and national security. Given that the first duty of any government is to secure the nation, this is unsurprising. In time of war, every effort must be made to secure victory as quickly and—in terms of military and civilian casualties—as cheaply as possible. That said, when war and pandemic coincide, it is not inconceivable that more will die from disease than will die on the battlefield. It is claimed that more US soldiers died from the Spanish flu than were killed on the Western Front. During the Pacific campaign of the Second World War, disease wreaked havoc amongst Allied forces.

The death and suffering seen in Philadelphia and Pittsburgh during the pandemic was attributable not only to those cities' prioritisation of the war effort, but also to chronic under-investment in basic infrastructure and the normalisation of environmental disamenity. Latent errors (weaknesses) (see [42, 43] for a definition) included:

- In Philadelphia, a transport network with insufficient capacity to cope with an influx of workers

- In Philadelphia and Pittsburgh, a housing sector with insufficient capacity and amenity to accommodate workers in reasonably sanitary conditions
- In Pittsburgh, a health-care system with insufficient capacity and expertise to meet the health challenges of a pandemic
- In Pittsburgh, an industrial sector whose emissions polluted the city and its environs and caused disease. Civic leaders tolerated this disamenity because heavy industry was beneficial to the city's economy. City politics were dominated by vested interests.

2.2.1 Lessons Learned and Actioned?

To what degree have governments facing the COVID-19 pandemic heeded the lessons of the Spanish flu pandemic? By June 9, 2020, the United Kingdom had recorded the second highest number of deaths in the world from COVID-19 [50]. The Chinese city of Wuhan, the epicentre of the pandemic, began its lockdown on 23 January, 2020. In Italy, there were selective lockdowns from 8 March, 2020. At this time there were frequent flights between the EU and the UK. In Britain, the Cheltenham Festival, a four-day-long horse racing event worth an estimated £100 million to the local economy [51], commenced on 10 March. An estimated 150,000 people attended over four days [52]. At the time the festival was held there were no social distancing measures in force. Racegoers massed in every corner of the facility. In early March the Prime Minister, Boris Johnson, set the tone for the country's encounter with COVID-19 when he proclaimed that Britons should "as far as possible, go about business as usual" (Johnson cited in [53]). Just before the festival commenced, the Culture Secretary, Oliver Dowden MP, heeded his leader's words. Responding to calls to cancel the event, he said: "There's no reason for people not to attend such events or to cancel them at this stage" (Dowden cited in [53]). While the United Kingdom charged ahead with a hectic schedule of prestigious and lucrative sporting fixtures, other countries were cancelling events: "On the first weekend in March, there was a full programme of football in both England and Scotland, five horse racing meetings, and Six Nations rugby at Twickenham between England and Wales—which the prime minister himself attended. It was a different matter elsewhere. A forthcoming Six Nations match in Dublin had already been postponed, along with the Chinese Grand Prix and football matches in virus-stricken northern Italy" [53].

Towards the end of April 2020, a former director of Public Health for the county in which the Cheltenham Festival was held, Professor Gabriel Scally, spoke out: "I think it's very tempting to link [the seeming high number of COVID-19 cases in Gloucestershire] to the Cheltenham Festival …. Really, from a health point of view, [it] should have been stopped in advance" (Scally cited in [52]). Towards the end of May, 2020, Professor Timothy Spector (cited in [53]) of King's College London (KCL) said: "I think sporting events should have been shut down at least a week earlier because they'll have caused increased suffering and death that wouldn't otherwise have occurred". Spector's comments referenced a COVID-19 hotspot in Gloucestershire.

A charitable characterisation of US President Donald Trump's behaviour during the early stages of the COVID-19 pandemic would be that it was uneven. According to Edward Luce of the Financial Times, over a few days, Trump shifted from denial to accepting that the US was not immune to pandemic. Then, during a visit to the Centres for Disease Control and Prevention (CDC), he publicly supported the CDC's work. Luce [54] observes: "Shortly before the CDC visit [on 6 March, 2020], Trump said 'within a couple of days, [infections are] going to be down to close to zero' [sic]. The US then had 15 cases [of COVID-19]. 'One day, it's like a miracle, it will disappear'. A few days afterwards, he claimed: 'I've felt it was a pandemic long before it was called a pandemic'. That afternoon at the CDC provides an X-ray into Trump's mind at the halfway point between denial and acceptance".

In a national emergency, leaders are obligated to act in the national interest, rather than in their own interest. An uncharitable characterisation of Trump's behaviour during the early stages of the COVID-19 pandemic would be that his decision-making was heavily influenced by his desire to keep the US economy afloat and the markets sweet (thereby maximising his chances of re-election in November 2020). Donald Trump's chances of re-election rested heavily on the health of the US economy. Aware of the damage a total lockdown could do to the economy, Trump dragged his feet. In public and private he boasted he had acted in a timely manner. Some claim he did too little, too late: "The psychology behind Trump's inaction on Covid-19 was on display … at the CDC. The unemployment number had come out that morning. The US had added 273,000 jobs in February, bringing the jobless rate down to a near record low of 3.5 per cent. Trump's re-election chances were looking 50:50 or better …. Nothing could be allowed to frighten the Dow Jones" [54].

According to some seasoned observers, Trump's behaviour—and that of his narrow policy-making coterie—during the pandemic and other crises, such as the civil unrest that followed the death of George Floyd at the hands of Minneapolis police officers [55], evidenced:

- Denialism
- Groundless optimism
- Ignorance
- Quackery
- Philistinism
- Naivety
- Mendacity
- Xenophobia
- Blamism
- Bullying
- Authoritarianism
- Vindictiveness
- Divisiveness
- Egocentrism
- Narcissism
- Indifference

- Partisanship
- Self-interestedness [54, 56–65].

Regarding claims of quackery, in late April, Trump publicly claimed that ingesting disinfectants, such as Reckitt Benckiser's Lysol or Dettol, might provide a curative treatment for COVID-19: "I see the disinfectant where it knocks it out in a minute, one minute .... Is there a way we can do something like that, by injection inside or almost a cleaning? Because you see it gets in the lungs, and it does a tremendous number on the lungs [sic], so it'd be interesting to check that" (Trump cited in [64]). (Note the rambling inarticulacy of his address). Disturbed by Trump's claims, Reckitt Benckiser issued a strongly-worded statement warning consumers not to ingest their products: "Due to recent speculation and social media activity, Reckitt Benckiser (the makers of Lysol and Dettol) has been asked whether internal administration of disinfectants may be appropriate for investigation or use as a treatment for coronavirus (SARS-CoV-2). As a global leader in health and hygiene products, we must be clear that under no circumstance should our disinfectant products be administered into the human body (through injection, ingestion or any other route)" (Reckitt Benckiser cited in [64]). It is claimed that Rex Tillerson, Trump's sacked Secretary of State, once referred to the President as a "moron" [65, 66] and that his Chief of Staff, John Kelly, once called Trump an "idiot" [65].

Regarding claims of indifference, at a political rally in Tulsa on 20 June 2020, Trump claimed he had asked that testing for COVID-19 be slowed. Trump's claim drew a warm response from the six thousand or so present (in a venue that can accommodate 19,000 people), a hostile response from others, a clarification from the White House and a statement of intent from one of the co-ordinators of the COVID-19 response, Dr Anthony Fauci (Fig. 7).

It is fair to say that Trump's claim—one reading of which was that he considered testing for COVID-19 to be politically inconvenient—drew significant criticism. Howard Forman, Professor of Public Health at Yale University, observed: "Our President didn't do the right thing in January. He didn't try in February. He seemed clueless in March. In April, thousands [and] thousands were dying weekly. In May he admitted he didn't like testing. Last night, he made it even more clear. Blood is on his hands" (Forman cited in [67]). Neurosurgeon and medical reporter Dr Sanjay Gupta observed: "There really is no correlation, in the President describing it, between an increase in testing and an increase in cases. Of course, you're going to find more cases .... The whole reason you do more testing is to find people, isolate them, slow down transmission and decrease the number of cases. If you're doing testing right, that is how it should work. We have not been doing testing right .... That is the one tool we ... had to potentially curb this pandemic .... What do ... other countries have? They have testing". Gupta described Trump's Tulsa claim that he had asked that testing for COVID-19 be slowed down a "public health travesty" (Gupta cited in [67]). Dr. Ashish Jha, Director of the Harvard Global Health Institute, remarked of Trump's claim that he had asked officials to slow testing: "It's actually very consistent with the policy we've been seeing coming out of the White House .... An effort to not put too much time and effort into ramping-up testing. And this is incredibly frustrating

The late June, 2020, exchanges over COVID-19 testing in the USA				
Date	20 June	20 June	21 June	23 June	23 June
Actor	Trump	White House	Navarro	Trump	Fauci
Comments	At a campaign rally in Tulsa, Oklahoma, Trump claimed he had asked officials to slow testing for COVID-19: "Testing is a double-edged sword .... Here's the bad part: when you do testing to that extent, you're going to find more people, you're going to find more cases. So I said to my people, slow the testing down please" (Trump cited in [67])	The White House claimed Trump's comments were made in jest. One official told CNN that the President was "obviously kidding" (unnamed source cited in [92]) Clearly, some insiders considered the President's claim potentially damaging	Appearing on CNN, Peter Navarro, a Trump advisor, said: "Come on now, that was tongue in cheek .... That was a light moment for him at a rally" (Navarro cited in [92])	The President insisted to reporters that his Tulsa comment was not intended as a joke. "I don't kid" he said (Trump cited in [73])	Addressing a Congressional committee, Dr Anthony Fauci, an infectious disease expert and a key player in efforts to tackle the pandemic, said: "[T]o my knowledge, none of us have ever been told to slow down on testing. That just is a fact. In fact, we will be doing more testing" (Fauci cited in [91])

Fig. 7 The political fall-out from Donald Trump's 20 June, 2020 claim about COVID-19 testing

for the millions of Americans who have gotten sick and have not been able to get tests. It's got to be incredibly frustrating for people who have lost family members in nursing homes .... [M]ore than 100,000 Americans [have] died largely because we have not built up the testing infrastructure that our country needs" (Jha cited in [68]). Kate Bedingfield, Democratic Senator Joe Biden's Deputy Campaign Manager and Communications Director, observed: "In an outrageous moment that will be remembered long after tonight's debacle of a rally, President Trump just admitted that he's putting politics ahead of the safety ... of the American people .... He ordered his people to 'slow the testing down', because he believed the large numbers were making him look bad. People have died because he failed to provide tests" (Bedingfield cited in [67]). Symone Sanders, an advisor to Joe Biden, remarked; "This is an appalling attempt to lessen the numbers only to make them look good .... That's what will be remembered long after last night's debacle of a rally—the admission of the President that he slowed testing for his political benefit" (Sanders cited in [69]).

Some weeks before the Tulsa rally, Trump had remarked: "[D]on't forget, we have more cases than anybody in the world. But why? Because we do more testing"

(Trump cited in [70]). Trump seemed ignorant of the fact that there is no correlation between the number of tests performed for COVID-19 and the number of cases of COVID-19. Tests do not generate cases. Tests are simply a reasonable indicator of the magnitude of the problem and a means of identifying those who should self-isolate. Testing reveals.

Finally, the Tulsa episode adds weight to the allegation that the Trump administration is chaotic and uncoordinated [71, 72]. On 20 June in Tulsa, Trump claimed that he had asked his staffers to "slow the testing down please" (Trump cited in [67]). Subsequently, the White House claimed the President's remark was tongue-in-cheek and had been made in jest. Reacting to the White House's assertion, Trump (cited in [73]) told reporters: "I don't kid".

It has been alleged that Trump brooks no dissent. Trump's behaviour has attracted comment from academics and former employees. Professor of Psychology and Human Development, Dan P. McAdams [60], observes: "Trump leads ... through intimidation, bluster and threat .... Trump's leadership style derives readily from his personality makeup, which entails a combustible temperament mixture of high extraversion and low agreeableness [and] a motivational agenda centred on extreme narcissism...". Authoritarian leadership stymies critique—a useful stress-test for policy. To use a term that has become fashionable, authoritarianism has a chilling effect. "The way to keep your job is to out-loyal everyone else, which means you have to tolerate quackery.... You have to flatter him [Trump] in public and flatter him in private. Above all, you must never make him feel ignorant" observes Anthony Scaramucci, Trump's former White House head of communications (Scaramucci cited in [54]).

Authoritarianism is a vector for groupthink, a situation where the members of an ideologically-coherent and socially-narrow group comprehend, reason and act in the same way. Groupthink produces ineffective and/or risk-laden analyses and decisions [19, 56, 74–76]. Psychologist Dr. Shawn Burn [56] opines: "[Trump] could benefit from learning more about issues, listening to credible sources and relying less on his gut and those that flatter him". According to Rhodes [77], President Donald Trump, apparently ignorant of the risks inherent in groupthink, has surrounded himself with people of like mind: "Instead of seeing U.S. government expertise as a resource, Trump has routinely derided career experts as 'deep state' operatives, insufficiently loyal to him and his agenda". According to Jurecic and Wittes [78], in early 2020, Trump moved against intelligence officials who, in his opinion, were producing analyses inconvenient to his political agenda. Trump replaced the acting Director of National Intelligence (DNI) with someone he knew to be loyal to him. The new DNI immediately removed the second in command at the Office of the Director of National Intelligence. It should be noted that in conducting this purge—according to Jurecic and Wittes [78], in a fit of pique—Trump did nothing illegal: the Constitution of the United States permits the President to seed the civil service with persons sympathetic to his/her agenda. Unfortunately, the power to remove civil servants whose professional analyses are deemed politically inconvenient opens the door to groupthink: "[T]he president spent [a] week reshaping the intelligence community to serve his political needs .... Removing intelligence officers for having the temerity to give their unvarnished assessments is a recipe for groupthink and analytic distortions based on what people think the boss wants to hear. Remember, these are the people the President needs to rely on when he decides whether to take military action. Do we really want them worrying about whether he will like their honest assessment, about whether telling the truth to Congress will give the opposition party information to use against the boss during an election season, or about whether they can blow the whistle on political lies about intelligence without retribution?" [78]. Proscribing political appointments would reduce the chances of groupthink within the various branches of an administration.

According to psychiatrist Dr. Tennyson Lee [79], a person with Narcissistic Personality Disorder (NPD) "Has a grandiose sense of self-importance .... Is preoccupied with fantasies of unlimited success, power, brilliance.... Believes that he or she is 'special'.... Requires excessive admiration [and] [s]hows arrogant, haughty behaviours or attitudes". Associate Professor of Psychology at St. John's University, Dr Aubrey Immelman [58], observes: "[T]he Trump presidency personifies a perilous combination of sparse political experience and the potential for a level of impulsiveness and hubris rarely—possibly never before—seen in occupants of the Oval Office". Psychology Today [61] observes: "[A] narrow area of agreement for most Americans is that the political climate has never been more corrosive, and that it reflects, to a greater or lesser degree, Trump's contrarian approach to leadership". The final word on Trump's pandemic performance goes to Gregg Gonsalves, a public health expert at Yale University: "Trump could have prevented mass deaths, and he didn't" (Gonsalves cited in [54]).

It is informative to contrast Trump's chaotic and authoritarian leadership style with that of St. Louis's Health Commissioner, Dr. Max C. Starkloff. By common agreement, Starkloff, while determined to save as many lives as possible during the Spanish flu pandemic, consulted widely and listened intently. He listened not only to doctors and epidemiologists, but also to those with vested interests, such as the city's business and religious leaders: "On October 7 [1918], Starkloff called Mayor Henry Kiel, representatives from the United States Public Health Service, the Red Cross, the St. Louis medical community, business interests, city hospitals and the public school system to his office to discuss the most effective way to fight the city's nascent epidemic. Several present were against the prospect of mass closures. With over one hundred civilian cases in the city and 900 cases at Jefferson Barracks, however, Starkloff urged them to consider drastic action in order to halt the spread of the disease. After some debate, the group agreed. They conferred Starkloff with legal authority to make public health edicts. They also agreed to a sweeping closure order" [46]. Starkloff's inclusive leadership style (a feature of high-reliability organising) helped save lives.

2.3   Modernity Delivering Ever More Comfort and Security?

For Beck [80, p. 50], Modernity consists of "surges of technological rationalization and changes in work and organisation". Modernity has delivered the Risk Society, in which environmental and other safety concerns are foregrounded: "[W]hile in classical industrial society, the 'logic' of wealth production dominates the 'logic' of risk production, in the risk society this relationship is reversed .... In the welfare states of the West … the struggle for one's 'daily bread' has lost its urgency …. Parallel to that the knowledge is spreading that the sources of wealth are 'polluted' by growing 'hazardous side-effects'" [80, pp. 12–20]. Beck coins the term 'boomerang effect' to encapsulate how some of the products of technological innovation, such as nuclear power and agrichemicals, can, under certain conditions, harm. In 1986, a concatenation of technological, procedural and cultural failures led to the Chernobyl nuclear disaster that deposited radioactive contamination over much of western Europe [81]. In the 1980s, Britain's Advisory Committee on Pesticides (ACP) claimed the defoliant 2,4,5-T was safe if used as directed. The National Union of Agricultural and Allied Workers argued that the ACP's advice on how the defoliant should be used was predicated on a model of farmworking that did not, in reality, exist, and that, as a consequence, 2,4,5-T posed a threat to farmworkers' health [82]. Boomerang effects occur " … where mostly unplanned results of (production) processes in modern societies backfire on these societies" [83].

Air service is one of Modernity's great successes. It unites. It creates opportunity and wealth [19, 84]. Unfortunately, it also facilitates the spread of disease: First, by providing affordable air service to almost every corner of the earth; Secondly, by using an environmental control system that circulates air within the aircraft cabin. Although filtered [85], the recirculated air provides a vector for distributing viruses to passengers and crew. The COVID-19 virus was seeded across the globe chiefly by aircraft passengers and crew—a classic boomerang effect wherein a triumph of modernity is transformed by circumstance and character into a vector for death and suffering. It could have been worse. Had the virus been EVD or something even more lethal, untold numbers could have died. The fatality rate for EVD varies between 25 and 90%. That for COVID-19 varies between 3 and 4% [86]. That for seasonal influenza is below 0.1%. The COVID-19 pandemic suggests that Modernity is Janus-faced, simultaneously delivering comfort and death (Fig. 8).

3   Conclusions

Politicians' decisions and actions shape pandemics. They influence where and how the pandemic proceeds. They influence who dies. During the COVID-19 pandemic, the UK government failed to ensure that care homes were adequately protected. The National Health Service (NHS)—relentlessly fetishised by politicians and the public—knowingly discharged untested elderly patients back into care homes,

Fig. 8 The two faces of nuclear fission. In 1945, the port city of Nagasaki was eviscerated with a nuclear device. In 1954, Lewis Strauss, chairman of the United States Atomic Energy Commission, promised that nuclear power stations would produce electricity "too cheap to meter". Fission is Janus-faced. [87]

causing death and suffering from COVID-19 [88]. In (at least) one case, the medical record of an elderly patient was altered by a National Health Service employee, from indicating the patient had tested positive for COVID-19 to indicating the patient had tested negative. The NHS claimed it had been unable to identify the employee who had altered the record without signing it (a breach of protocol) [89].

Politicians' self-interested behaviour and denialism meant that the pandemic took root. In the UK, the Johnson administration claimed its decisions were 'led by the science', thereby providing it with a scapegoat—the scientific community. In reality, of course, the Johnson administration's decisions were only informed by the science. In the final analysis, decisions are made by politicians alone. Other influences on COVID-19 policy included a determination to protect the financial interests of the hospitality and manufacturing sectors. Denying responsibility for contentious decisions provides politicians with a safeguard against being held to account.

When the COVID-19 threat emerged in 2019, politicians could have learned lessons from the good and bad practices seen during the 1918–1919 Spanish flu pandemic. For example, from St.Louis's and New York's civic-minded proaction in the face of a serious threat to public health [46, 90]. Application of these lessons through active learning [20] might have saved lives. Unfortunately, active learning is never guaranteed [21, 22]. Perforce, politicians balance interests. Public health is but one interest amongst many. Interests balanced during the Spanish flu pandemic included public health, corporate profits and war aims.

Indisputably, Modernity has delivered a higher standard of living: "Poverty, malnutrition, illiteracy, child labour and infant mortality are falling faster than at any other time in human history. Life expectancy at birth has increased more than twice as much in the past century as it did in the previous 200,000 years" [1, p. 31]. Thanks to the boomerang effect, however, as quickly as Modernity mitigates or eliminates risks, new risks—such as the chemical contamination of groundwater, pollution of the oceans and food chain with microplastics, global warming, theft of large volumes of personal data from corporate datastores or viruses spread with deadly efficiency by airline passengers and crew—fill the void. Sometimes, progress seems like a zero-sum game.

References

1. Norberg J (2016) Things have only got better. The Sunday Times, 11 September
2. Husain M (2013) Malala: The girl who was shot for going to school. Available at: https://www.bbc.co.uk/news/magazine-24379018/. Accessed 20 May 2020
3. Yousafzai M (2009) Diary of a Pakistani schoolgirl (i). Available at: https://news.bbc.co.uk/1/hi/world/south_asia/7834402.stm/. Accessed 20 May 2020
4. Yousafzai M (2013) I am Malala. Weidenfeld and Nicolson, London
5. Arbabzadah N (2011) The 1980s Mujahideen, the Taliban and the shifting idea of jihad. The Guardian, 28 April
6. Crile G (2007) Charlie Wilson's War. Grove Atlantic, New York NY
7. Clarke P (2014) Report into allegations concerning Birmingham schools arising from the 'Trojan horse' letter (HC 576). Department for Education, London
8. Wikimedia Commons (2015) Portrait of President Roosevelt. Available at: https://commons.wikimedia.org/wiki/File:Cropped_Portrait_of_FDR.jpg/. Accessed 23 May 2020
9. National Health Service (2020a) Polio. Available at: https://www.nhs.uk/conditions/polio/. Accessed 21 May 2020
10. Baicus A (2012) History of polio vaccination. World J Virol 1(4):108–114
11. World Health Organisation (2017) 10 facts on polio eradication. Available at: 2020 https://www.who.int/features/factfiles/polio/en/. Accessed 20 May 2020
12. Toft B, Reynolds S (2005) Learning from disasters: A Management approach. Palgrave-Macmillan, Basingstoke
13. World Health Organisation (2020a) Ebola virus disease. Available at: https://www.who.int/news-room/fact-sheets/detail/ebola-virus-disease/. Accessed 21 May 2020
14. National Health Service (2020b) SARS (severe acute respiratory syndrome). Available at: https://www.nhs.uk/conditions/sars/. Accessed 21 May 2020
15. The Visual and Data Journalism Team (2020) Coronavirus pandemic: tracking the global outbreak. Available at: https://www.bbc.co.uk/news/world-51235105/. Accessed 20 May 2020
16. Gopinath G (2020) The great lockdown: worst economic downturn since the great depression. Available at: https://blogs.imf.org/2020/04/14/the-great-lockdown-worst-economic-downturn-since-the-great-depression/. Accessed 20 May 2020
17. Partington R (2020) Chancellor plays down hopes of quick economic recovery. Rishi Sunak warns UK facing recession 'the likes of which we haven't seen' due to Covid-19 crisis. Available at: https://www.theguardian.com/business/2020/may/19/chancellor-plays-down-hopes-of-quick-economic-recovery/. Accessed 21 May 2020

18. Sweney M (2020) Rolls-Royce to cut 9,000 jobs as Covid-19 takes toll on airlines. Available at: https://www.theguardian.com/business/2020/may/20/rolls-royce-to-cut-9000-jobs-worldw ide-as-covid-19-takes-toll/. Accessed 21 May 2020
19. Bennett SA (2019) Systems-thinking for Safety. A short introduction to the theory and practice of systems-thinking. Peter Lang International Academic Publishers, Oxford
20. Toft B (1992) The failure of hindsight. Disaster Prev Manage: Int J 1(3):48–63
21. Bennett SA (1999) Learning the lessons of the past—opportunities and obstacles. Proceedings, Innovative Technologies for Disaster Mitigation: An architectural surety conference. Washington DC, 27–30 October
22. Bennett SA (2001) Case Studies in Architectural Surety. Disaster research report, vol. 2, No. 1. Institute of Civil Defence and Disaster Studies, Worcester
23. Buchanan DA (2011) Reflections: Good practice, not rocket science—understanding failures to change after extreme events. J Chang Manag 11(3):273–288
24. Barry JM (2004) The Great Influenza. Penguin Publishing, New York NY
25. Barry JM (2017) How the horrific 1918 flu spread across America. Smithsonian Magazine, November
26. Children's Hospital of Philadelphia (2020) Vaccine history: developments by year. Available at: https://www.chop.edu/centers-programs/vaccine-education-center/vaccine-history/develo pments-by-year#/. Accessed 8 June 2020
27. History.com (2020) Spanish Flu. Available at: https://www.history.com/topics/world-war-i/ 1918-flu-pandemic/. Accessed 22 May 2020
28. Wikimedia Commons (2020b) Policemen in Seattle wearing masks made by the Red Cross, during the influenza epidemic. December 1918. Available at: https://commons.wikimedia.org/ wiki/File:Spanish_flu_in_1918,_Police_officers_in_masks,_Seattle_Police_Department_deta il,_from-_165-WW-269B-25-police-l_(cropped).jpg/. Accessed 23 May 2020
29. White KA (1985) Pittsburgh in the great epidemic of 1918. Western Pennsylvania Hist Mag 68(3):221–242
30. Wikimedia Commons (2020a) Camp Funston, at Fort Riley, Kansas, during the 1918 Spanish flu pandemic. Available at: https://commons.wikimedia.org/wiki/File:Camp_Funston,_at_Fort_R iley,_Kansas,_during_the_1918_Spanish_flu_pandemic.jpg/. Accessed 23 May 2020
31. Zeitz J (2020) Rampant lies, fake cures and not enough beds: What the Spanish flu debacle can teach us about coronavirus. Available at: https://www.politico.com/news/magazine/2020/ 03/17/spanish-flu-lessons-coronavirus-133888/. Accessed 22 May 2020
32. Reporters Without Borders (2020) China. Available at: https://rsf.org/en/china/. Accessed 24 May 2020
33. Xu B, Albert E (2017) Council on Foreign Relations: Media censorship in China. Available at: https://www.cfr.org/backgrounder/media-censorship-china/. Accessed 24 May 2020
34. Williams-Herrman E (2013) Pittsburgh in World War I: Arsenal of the Allies. History Press (SC), Charleston SC
35. Daniels M (2014) The Great War and the Steel City. Available at: https://archive.triblive.com/ news/the-great-war-and-the-steel-city/. Accessed 26 May 2020
36. Wikimedia Commons (2018) Mills in Strip District, Pittsburgh. View of Pittsburgh's Strip District neighborhood looking northwest from the roof of Union Station. The Fort Wayne railroad bridge spanning the Allegheny River is partially visible on the left. Available at: https://com mons.wikimedia.org/wiki/File:Mills_in_Strip_District,_Pittsburgh_(84.41.70).jpg/. Accessed 26 May 2020
37. O'Toole W (2020) When the Spanish flu swept in, Pittsburgh failed the test. Available at: https://pittsburghquarterly.com/articles/when-the-spanish-flu-swept-in-pittsburgh-fai led-the-test/. Accessed 24 May 2020
38. University of Michigan Centre for the History of Medicine (2020a) Influenza encyclopaedia. Pittsburgh, Pennsylvania. Available at: 2020 https://www.influenzaarchive.org/cities/city-pit tsburgh.html/. Accessed 24 May 2020
39. Butler SD (1935) War is a Racket. Round Table Press, New York NY

40. Military-Industrial Complex (2020) What is the military-industrial complex? Available at: https://www.militaryindustrialcomplex.com/what-is-the-military-industrial-complex.asp/. Accessed 8 June 2020

41. Irving H (2014) Chaos and Censorship in the Second World War. Available at: https://history.blog.gov.uk/2014/09/12/chaos-and-censorship/. Accessed 25 May 2020

42. Reason JT (1990) Human error. Cambridge University Press, Cambridge

43. Reason JT (2013) A Life in error. Ashgate, Farnham

44. Allie SP (2018) St. Louis and the Great War. The University of Chicago Press, Chicago Il

45. Leonard MD (2017) 100 years later, St. Louis and the nation remember 'The Great War'. Available at: https://news.stlpublicradio.org/post/100-years-later-st-louis-and-nation-remember-great-war/. Accessed 27 May 2020

46. University of Michigan Centre for the History of Medicine (2020b) Influenza encyclopaedia. St. Louis, Missouri. Available at: 2020 https://www.influenzaarchive.org/cities/city-stlouis.html/. Accessed 27 May 2020

47. Mason RO (2004) Lessons in organisational ethics from the Columbia disaster: Can a culture be lethal? Organisational Dynam 33(2):128–142

48. Roberts KH (1990) Some characteristics of one type of high-reliability organisation. Organisation Sci 1(2):160–176

49. Bennett SA (2012) Aviation and Corporate Social Responsibility. In: Innovative Thinking in Risk, Crisis and Disaster Management. Gower, Farnham, pp 139–171

50. Krever M, Rahim Z (2020) UK coronavirus death toll tops 50,000, new data shows. Available at: https://edition.cnn.com/2020/06/09/uk-coronavirus-deaths-ons-intl/index.html/. Accessed 9 June 2020

51. Reeves M (2019) The festival at Cheltenham Racecourse boosts the local economy by £100 million. Available at: https://www.gloucestershirelive.co.uk/news/business/festival-cheltenham-racecourse-boosts-local-2495302/. Accessed 29 May 2020

52. British Broadcasting Corporation (2020). Coronavirus: Calls for inquiry on Cheltenham Festival go-ahead. Available at: 2020 https://www.bbc.co.uk/news/uk-england-gloucestershire-52380282/. Accessed 29 May 2020

53. Tucker M, Goldberg A (2020) Coronavirus: Sports events in March 'caused increased suffering and death'. Available at: https://www.bbc.co.uk/news/uk-52797002/. Accessed 29 May 2020

54. Luce E (2020) FT Magazine. Coronavirus. Inside Trump's Coronavirus Meltdown. Available at: 2020 https://www.ft.com/content/97dc7de6-940b-11ea-abcd-371e24b679ed/. Accessed 29 May 2020.

55. Goldberg J (2020) James Mattis denounces President Trump, describes him as a Threat to the Constitution. https://www.theatlantic.com/politics/archive/2020/06/james-mattis-denounces-trump-protests-militarization/612640/. Accessed 4 June 2020

56. Burn SM (2019) The perils of Trump's autocratic leadership style. Available at: https://www.psychologytoday.com/gb/blog/presence-mind/201901/the-perils-trump-s-autocratic-leadership-style/. Accessed 31 May 2020

57. Goodman R, Schulkin D (2020) Timeline of the Coronavirus pandemic and U.S. response. Available at: https://www.justsecurity.org/69650/timeline-of-the-coronavirus-pandemic-and-u-s-response/. Accessed 29 May 2020

58. Immelman A (2017) The leadership style of U.S. President Donald J. Trump. (Working Paper No. 1.2). Collegeville and St. Joseph, MN: St. John's University and the College of St. Benedict, Unit for the Study of Personality in Politics. Available at: https://digitalcommons.csbsju.edu/psychology_pubs/107/. Accessed 31 May 2020

59. Mattis JN (2020) In union there is strength. Available at: https://edition.cnn.com/2020/06/03/politics/mattis-protests-statement/index.html/. Accessed 4 June 2020

60. McAdams DP (2017) The appeal of the primal leader: Human evolution and Donald J Trump. Evol Stud Imaginative Cult 1(2):1–13

61. Psychology Today (2020) Why is the Trump presidency of extreme psychological interest? Available at: https://www.psychologytoday.com/gb/basics/president-donald-trump/. Accessed 31 May 2020

62. Rim C (2019) Here's why Donald Trump doesn't want anyone to know his grades or SAT Scores. Available at: https://www.forbes.com/sites/christopherrim/2019/02/28/heres-why-donald-trump-doesnt-want-anyone-to-know-his-grades-or-sat-scores/#48a2ac313764/. Accessed 3 June 2020

63. The Guardian (2020) The Guardian view on Trump and Covid-19: Americans suffer. Will he? Available at: https://www.theguardian.com/commentisfree/2020/may/01/the-guardian-view-on-trump-and-covid-19-americans-suffer-will-he/. Accessed 29 May 2020

64. Togoh I (2020) Dettol and Lysol maker warns against drinking and injecting disinfectant after Trump suggestion. Available at: https://www.forbes.com/sites/isabeltogoh/2020/04/24/dettol-and-lysol-maker-warns-against-drinking-and-injecting-disinfectant-after-trump-suggestion/#7b99469346a5/. Accessed 29 May 2020

65. Weiland N (2018) 5 takeaways from Bob Woodward's book on the Trump Whitehouse. Available at: https://www.nytimes.com/2018/09/04/us/politics/trump-woodward-book-fear.html/. Accessed 25 June 2020

66. British Broadcasting Corporation (2018) Rex Tillerson—Trump's former top diplomat. It was no secret in Washington that former US Secretary of State Rex Tillerson and President Donald Trump had a difficult working relationship before Mr Trump fired him. Available at: 2020 https://www.bbc.co.uk/news/world-us-canada-38281954/. Accessed 14 March 2021

67. Zoellner D (2020) 'Blood on his hands': Trump under fire for claim that he ordered slow-down in coronavirus testing. Available at: https://www.msn.com/en-sg/news/world/e2-80-98blood-on-his-hands-e2-80-99-trump-under-fire-for-claim-that-he-ordered-slow-down-in-coronavirus-testing/ar-BB15Nyjq/. Accessed 24 June 2020

68. Dwyer D (2020) 'This is unfortunately not a joke': Harvard doctor reacts to Trump's comments on COVID-19 testing at Tulsa rally. Available at: https://www.boston.com/news/coronavirus/2020/06/22/ashish-jha-trump-rally-slow-down-testing/. Accessed 25 June 2020

69. Collinson S (2020) Trump's 'kidding' on testing exposes his negligence as virus spikes. Available at: https://edition.cnn.com/2020/06/22/politics/donald-trump-coronavirus-testing-tulsa-rally/index.html/. Accessed 25 June 2020

70. Lee BY (2020) Trump: Without Coronavirus testing 'We would have very few cases.' Here is the reaction. Available at: https://www.forbes.com/sites/brucelee/2020/05/15/trump-without-doing-covid-19-coronavirus-testing-we-would-have-very-few-cases-here-is-the-reaction/#7730ff2b518c/. Accessed 25 June 2020

71. Rucker P, Leonnig C (2020) A Very Stable Genius: Donald J. Trump's testing of America. Bloomsbury, London

72. Woodward R (2018) Fear: Trump in the White House. Simon and Schuster, New York NY

73. Gittleson B (2020) Trump says 'I don't kid' after aides argue he was joking about slowing Coronavirus testing. His deputies had said the President had been joking at a recent rally. Available at: https://abcnews.go.com/Politics/trump-kid-aides-argue-joking-slowing-coronavirus-testing/story?id=71404943/. Accessed 24 June 2020

74. Bennett SA (2019b) The 2018 Gosport Independent Panel report into deaths at the National Health Service's Gosport War Memorial Hospital. Does the culture of the medical profession influence health outcomes? J Risk Res https://doi.org/10.1080/13669877.2019.1591488

75. Janis IL (1972) Victims of Groupthink: A psychological study of foreign-policy decisions and fiascos. Houghton Mifflin, Boston MA

76. Neck CP, Moorhead G (1995) Groupthink remodelled: The importance of leadership, time-pressure and methodical decision-making procedures. Human Relat 48(5):537–557

77. Rhodes B (2020) How Trump designed his White House to fail. Every President chooses how to manage the flow of information. The consequences of Trump's decisions are now becoming apparent. Available at: https://www.theatlantic.com/ideas/archive/2020/03/white-house-set-fail/607960/. Accessed 26 June 2020

78. Jurecic Q, Wittes B (2020) Trump's most dangerous destruction yet. What the President is doing to America's intelligence community could have enormous repercussions for the 2020 election and the country's preparedness for threats from around the world. Available at: https://www.theatlantic.com/ideas/archive/2020/02/trumps-most-dangerous-destruction-yet/607021/. Accessed 26 June 2020

79. Lee TC (2017) Does Donald Trump have a personality disorder? Available at: https://www.psy chiatry-uk.com/donald-trump-personality-disorder/. Accessed 31 May 2020
80. Beck U (1992) Risk Society. Towards a New Modernity. Sage Publications, London
81. Mackay L, Thompson M (1988) Something in the Wind: Politics after Chernobyl. Pluto Press, London
82. Irwin A (1995) Citizen Science. Routledge, London
83. Wimmer J, Quandt T (2006) Living in the Risk Society. J Stud 7(2):336–347. https://doi.org/10.1080/14616700600645461
84. Woolford S, Warner C (2010) The Story of Flight. A history of aviation. Sevenoaks, London
85. National Research Council (US) Committee on Air Quality in Passenger Cabins of Commercial Aircraft (2002) The airliner cabin environment and the health of passengers and crew. National Academies Press, Washington DC
86. World Health Organisation (2020b) Coronavirus disease 2019 (COVID-19) Situation Report—46. World Health Organisation, Geneva
87. Wikimedia Commons (2013) The atomic cloud over Nagasaki. Available at: https://commons.wikimedia.org/wiki/File:Nagasakibomb.jpg/. Accessed 3 June 2020
88. Lintern S (2020). Coronavirus: Care homes still expected to take Covid-19 hospital patients as deaths mount. Available at: https://www.independent.co.uk/news/health/coronavirus-care-homes-nhs-deaths-statistics-a9500326.html/. Accessed 3 June 2020
89. Martin N (2020) Coronavirus: NHS trust apologises to family of elderly patient whose medical records were altered. Available at: https://news.sky.com/story/coronavirus-nhs-trust-apolog ises-to-family-of-elderly-patient-whose-medical-records-were-altered-11992855/. Accessed 3 June 2020
90. Aimone F (2010) The 1918 influenza epidemic in New York City: A review of the public health response. Public Health Rep 125(3_suppl):71–79
91. Holpuch A (2020) Coronavirus has brought US 'to its knees', says CDC Director. Dr Robert Redfield tells hearing that public health capabilities underfunded as US sees more than 2.3 million cases. Available at: https://www.theguardian.com/world/2020/jun/23/anthony-fauci-covid-19-statement-house-hearing/. Accessed 24 June 2020
92. Reston M (2020) White House officials on the defensive after Trump says he wanted testing slowed down. Available at: 2020 https://edition.cnn.com/2020/06/20/politics/tulsa-rally-trump/index.html/. Accessed 24 June 2020

Empowering Citizens with Tools for Personalized Health is the Future of Effective Public Health Responses

Jordan Masys, Chris Peng, Andrew Ahn, and Anthony J. Masys

Abstract The global health security landscape is characterized as 'messes' and 'wicked problems' that proliferate in this age of complexity. In designing global health security solutions, '…simple, quick fix solutions that flounder in the face of interconnectedness, volatility and uncertainty' (Jackson in Critical Systems Thinking and the Management of Complexity. John Wiley and Sons, NJ, 2019 [1]) will not suffice and may create unintended consequences. For example, recent years have seen the threat presented by bio-risks heighten significantly. Infectious disease emergencies have included Severe Acute Respiratory Syndrome (SARS), Foot and Mouth Disease (FMD), Middle East Respiratory Syndrome (MERS), H1N1 influenza, Ebola, Zika and now COVID-19. These outbreaks have had significant global and public health impact. Considering the current COVID-19 pandemic, as of January 2021, the COVID-19 pandemic has seen upwards of 100 million cases, 2.1 million deaths, and has significantly impacted the global economy and affected the most vulnerable. The impacts have seen the closure of borders, economic disruptions and failures, strained and overwhelmed health care systems, failure of supply chains, all of which are contributing to a human and national security issue. Sensemaking is the activity that enables us to turn the ongoing complexity of the world into a 'situation that is comprehended explicitly in words and that serves as a springboard into action' (Weick et al. in Organ Sci 16:409–421, 2005 [2]). The importance of sensemaking as it pertains to individual and public health is that it enables us to plan, design and take action in real-time to events that affect or 'shock' our health and healthcare system. With advances in technology over the last decade, from robust cloud infrastructures, to artificial intelligence, to wearable devices, personalized health interventions are becoming increasingly accurate, accessible, and prevalent. As part of this shift, there has been a growing movement in the field of public health to include personalized medicine as an integrative element of public health which would have the potential to

J. Masys · C. Peng
KIIPO, Taipei City, Taiwan

A. Ahn
Center for Dynamical Biomarkers, Harvard Medical School, Boston, MA, USA

A. J. Masys (✉)
College of Public Health, University of South Florida, Tampa, FL, USA

radically transform its methods and character (Evangelatos et al. in Int J Public Health 63:433–434, 2018 [3]). This is changing the landscape as it pertains to health 'sense-making' at the individual and public health level. This chapter presents the research and operationalization of 'biomarker' wearable technology as a tool for personalized health to support effective public health responses. This is contextualized within the backdrop of the COVID-19 pandemic.

Keywords Biomarkers · COVID-19 · Wearable · AI · Analytics

1 Introduction

The global health security landscape is characterized by 'messes' and 'wicked problems' that proliferate in this age of complexity. In designing global health security solutions, '…simple, quick fix solutions that flounder in the face of interconnectedness, volatility and uncertainty' [1] will not suffice and may create unintended consequences. For example, recent years have seen the threat presented by bio-risks heighten significantly. Infectious disease emergencies have included Severe Acute Respiratory Syndrome (SARS), Foot and Mouth Disease (FMD), Middle East Respiratory Syndrome (MERS), H1N1 influenza, Ebola, Zika and now COVID-19. These outbreaks have had significant global and public health impact. As of January 2021, the COVID-19 pandemic has seen upwards of 100 million cases, 2.1 million deaths, and has significantly impacted the global economy and affected the most vulnerable. The impacts have seen the closure of borders, economic disruptions and failures, strained and overwhelmed health care systems, failure of supply chains all of which are contributing to a human and national security issue.

Sensemaking is the activity that enables us to turn the ongoing complexity of the world into a 'situation that is comprehended explicitly in words and that serves as a springboard into action' [2]. The importance of sensemaking as it pertains to public health is that it enables us to plan, design and act to 'shocks' to our health and healthcare system in real-time. It is argued in Ming et al. [3] that 'Clinical studies have consistently demonstrated the need for timely medical intervention; delayed recognition of illness results in adverse outcomes and increased costs'. With advances in technology over the last decade, from robust cloud infrastructures, to artificial intelligence, to wearable devices, personalized health interventions are becoming increasingly accurate, accessible, and prevalent. As part of this shift, there has been a growing movement in the field of public health to include personalized medicine as an integrative element of public health which would have the potential to radically transform its methods and character [4]. As discussed in Soliño-Fernandez et al. [5], wearables '…have the potential to improve population health by moving the focus from disease treatment to prevention; routinely monitoring personalized physiological measurements; supporting self-management; identifying alterations in health conditions; and creating positive long term behavioral changes towards healthy lifestyles'. With an obvious impact on the health of individuals, it would also lead to an increase in the quality of data [6] from which public health experts can shape their strategies—great data can allow for great strategies and outcomes, but poor data will guarantee poor strategies and outcomes [7].

Through COVID-19, we've seen that the collection of real-time data—and quickly making sense of that data- is paramount to providing quick and accurate public health responses [8].

This chapter presents the research and operationalization of 'biomarker' wearable technology as a tool for personalized health to support effective public health responses. In highlighting some examples of deployed personalized health technologies, this chapter looks to further inspire the public health sector to embrace these new technologies, both to protect and improve the health of people but also to empower clinicians, researchers, and public health experts with real-time, real-world data.

Sensemaking.

We have seen significant individual and public health challenges associated with such outbreaks as SARS (2003), H1N1 (2009), MERS (2011), and now COVID-19. In addition, across populations, health challenges associated with multiple chronic conditions, from diabetes, hypertension, and asthma to depression, chronic-pain, sleep and neurological disorders put a strain not only on individual health but on the healthcare system. The requirement to monitor and understand symptoms, side effects and treatment outside the clinical setting is a game-changer for individual health, population health/surveillance and the healthcare system in general. Having an awareness and understanding of individual health biomarkers 'in situ and in real-time' represents a 'sensemaking' capability that can guide self-care and healthcare actions. Scaling such data aggregation at the regional, national or global level can support public health strategy design and deployment.

Weick [9] refers to sensemaking in terms of '…how we structure the unknown so as to be able to act in it'. From a data centric perspective, Klein et al. [10] describes the role of sensemaking in terms of '…explaining anomalies, anticipating difficulties, detecting problems, guiding information search, and taking effective action'. This is a significant capability to support individual and public health design of intervention strategies.

2 Biomarkers, Wearable Technologies, and AI are Shaping the Future of Personalized Health

Background on Biomarkers

The National Institute of Health Biomarkers Definitions Working Group defined a biomarker in 1998 as "a characteristic that is objectively measured and evaluated as an indicator of normal biological processes, pathogenic processes or pharmacologic responses to a therapeutic intervention" [11]. Biomarkers span a wide range of measures from simple vital sign metrics such as blood pressure and temperature to more complex ones such as genetic signatures or histological findings. By its very definition, it is both objective and measurable, while also—in theory—providing an accurate window into the state of an individuals' health. However, this notion of biomarkers as a good proxy for health does not always hold true, and biomarkers are more frequently than not imperfect surrogates of one's condition. For instance, elevated Brain-Natriuretic Peptide (BNP) as defined as $> = 100$ ng/L has only a positive predictive value of 67% to diagnose the presence of congestive heart failure [12].

Over the years, the term "biomarker" has come to be understood in medicine as a test that is (1) singular, (2) predominantly biochemical in nature and (3) specific to a certain disease condition. For example, an elevated CA19-9 biomarker is suggestive of pancreatic cancer while a positive Quantiferon-TB Gold test is indicative of Mycobacterium tuberculosis infection. "Biomarkers" are not typically used to measure "good health", nor does it rely on mechanical or electrical measures of the physical body. This emphasis on a singular, biochemical test is a by-product of the prevailing paradigm in modern medicine where patients visit clinical sites (hospital or outpatient clinics) episodically—possibly once or twice a year—and where diseases are largely understood as disordered states arising from biochemical derangements.

With the advent of portable technologies and high-throughput tools, however, "biomarkers" are likely to take on a much more dynamic, complex, and nuanced definition. Rather than focusing on a single, momentary test, clinicians may begin evaluating the temporal dynamics of continuous measures such as blood glucose, heart rate, or temperature obtained from wearable devices. Moreover, the availability of mass spectrometry and microarray chips has made it feasible to obtain multiple measures simultaneously and thereby promote the practice of evaluating the full composite of measures to gain better insights into one's health. These measures obtained across time and multiple dimensions provide a much richer view into the biological state of a person and may lead to a health assessment that is more graded in magnitude rather than assume a disease vs. no-disease dichotomy.

Wearable technologies and AI

As described in Ming et al. [3] 'Optimal management of infectious diseases is guided by up-to-date information at the individual and public health levels. For infections of global importance, including emerging pandemics such as COVID-19 or prevalent endemic diseases such as dengue, identifying patients at risk of severe disease and clinical deterioration can be challenging, considering that the majority present with a mild illness'. Hence the move towards embracing wearable technologies and AI to support health management.

With the rapid improvement of wearable devices over the last decade and their rising popularity, the opportunity to collect continuous, real-world data (i.e. physiological, biological, environmental, behavioral data), is now more accessible than ever. Wearable technologies, from ECGs to wristbands with PPG sensors, are already being used for the detection and monitoring of various types of diseases, such as respiratory diseases, sleep diseases, cardiovascular diseases, etc. [13, 14]. For example, connected continuous glucose monitoring systems combined with closed-loop insulin delivery systems have been found to improve type 2 diabetes mellitus control [15]. We are also seeing these capabilities being deployed in consumer-grade devices, such as the Apple Watch, by using ECG waveforms to monitor for atrial fibrillation [16].

With advances in physiological signals analysis, "derived" biomarkers will continue to be discovered, further expanding the applicability of wearable devices. For example, Cardiopulmonary coupling (CPC) is an algorithm that can use ECG signals to derive respiration during sleep, and is used to quantify sleep quality and note instances of sleep apnea [17].

The utilization of AI and other machine learning technologies will be integral to the further advancement of health and medicine. AI can already outperform medical practitioners in the analysis of skin lesions, pathology slides, electrocardiograms or medical imaging data [18, 19]. The thousands of data points collected from wearable devices, coupled with the power of artificial intelligence, can then help to inform diagnosis, predict patient outcomes, and guide treatment decisions or proactive interventions [20]. Despite many of the rewarding aspects of wearable devices, one of its drawbacks is the 'noisy' or unsanitized data arising from real-world data collection. AI systems can help overcome this challenge by compensating for the noisy, artefactual signals [21].

The scalable application of wearable devices and AI and its impact on personalized health have been seen most recently through COVID-19. In the following section, we explore how personalized health tools have been used to address some of the public health challenges posed by COVID-19.

3 Deploying Personalized Health Technologies as a Public Health Strategy for COVID-19

(a) Empowering citizens with tools to protect and improve their health

The clinical spectrum of COVID-19 ranges from an asymptomatic or mild flu-like illness to a severe pneumonia requiring critical care [7]. Frequently reported signs and symptoms of patients admitted to the hospital include fever, cough, myalgia or fatigue, and shortness of breath at illness onset [22, 23], with a patient's full recovery potentially taking several weeks. With such a large influx of patients, hospitals are forced to discharge patients as soon as some improvement in symptoms are observed in order to keep up with the priority cases. In addition, mild or low-priority cases may no longer qualify for in-hospital monitoring [22], and at-home management becomes a necessary option provided they are able to safely self-isolate. However, patients require tools to self-monitor, the ability to get tested and contact their doctors in case of clinical deterioration. Easily-collectible biomarkers such as SpO2, body temperature, and heart rate, when monitored in real-time can provide both the patient and his/her physician with the data they need to take appropriate action [24].

According to the CDC, 50% of COVID-19 transmissions occur prior to an individual feeling sick or showing any symptoms [25]. The current limitation of the most common tools for symptom detection, such as surveys and temperature measurements, is that these methods are likely to miss pre-symptomatic or asymptomatic cases. For example, an elevated temperature (>100 °F (>37.8 °C)) is not as common as frequently believed, being present in only 12% of individuals who tested positive for COVID-19 and just 31% of patients hospitalized with COVID-19 (at the time of admission) [26]. It has become clear that we cannot simply rely on a single biomarker; the more COVID-19 biomarkers we collect, the higher our chances of early detection.

With advances in wearable technology, it is now possible to continuously monitor a multitude of biomarkers. As opposed to the typical subjective symptom input, wearable devices give us real objective data. A recent study published by the Scripps Research Translational Institute highlighted that sensor data from consumer-level wearable devices can help predict COVID-19 infection and can complement virus testing and conventional screening to signal new infections [26]. One of the greatest public health challenges in stopping COVID-19 from spreading is the ability to quickly identify, trace and isolate infected individuals. Early identification of those who are pre-symptomatic or even asymptomatic would be an invaluable tool for public health experts, as individuals may potentially be even more infectious during this period [27].

(b)   **Empowering clinicians, researchers and public health experts with the real-time, real-world data needed for effective strategies and responses**

Real-time, real-world data and for clinicians and researchers

Despite the tremendous amount of research having been done on COVID-19, research is predominantly concentrated in those patients being hospitalized (either on the floor or the ICU). This makes perfect sense because the tools needed to collect clinically relevant data are extensively located in the hospital: i.e., vitals monitoring (temperature, heart rate, blood pressure, respiratory rate, oxygen saturation), blood tests, imaging (chest x-ray, CT scan, ultrasound), EKG/telemetry, and physical exam. However, because of this, our knowledge about COVID-19 cases outside of the hospital is generally lacking—not only in times prior to hospitalization and after discharge, but also for individuals who never presented to the hospital/clinic in the first place due to very mild-to-moderate symptoms. There may also be those unfortunate cases where individuals suffered an untimely death due to COVID-19 without ever reaching a hospital.

Consequently, there are aspects of the disease that remained largely unknown for a long time: what is the most common initial symptom associated with COVID-19 (fatigue, sore throat, loss of smell, etc.)?; what are the most common symptoms seen in more mild-moderate cases and what are their time courses?; How do select population groups (children, elderly, those with diabetes, hypertension or on immunosuppressant medications, etc.) respond to SARS-CoV2 infection?; how are patients doing after discharge?; are they recovering slowly/quickly and does age,

COVID disease severity, or other comorbidities affect the way they recover?; and if discharged patients require oxygen, for what duration is oxygen supplementation required? For many clinicians, these questions continue to emerge as they accumulate experiences taking care of COVID-19 patients in varying scenarios. It has become increasingly obvious that major gaps in knowledge continue to exist [22].

Platforms that allow individuals to input their own symptoms/observations while also incorporating wearable data which can further corroborate or elaborate one's experience through this unprecedented crisis, represent a great opportunity to gain insights into these under-evaluated settings. This pairing between subjective data and objective information within a day-to-day, real-world setting and in a continuous fashion empowers the data-scientist/medical-expert to perform analyses not typically available in traditional datasets. Most health datasets rely on intermittent, episodic data collection—at weekly, monthly or yearly intervals—assuming they are even obtained on a regular basis at all. And for those studies that do, such as epidemiological studies (e.g., Framingham-cohort study [28]) or randomized controlled trials, data collection is done under exceptionally costly conditions due to intensive labor and supply requirements.

Wearable technology, in particular, has dramatically transformed our capabilities of addressing research questions. With it, we can potentially assess (1) the effects of sleep on our immunity/resistance to COVID-19 or the symptoms associated with it, (2) how stress (as measured by heart rate variability or galvanic skin response) may affect our COVID-related symptoms, (3) whether there are specific times in the day when fevers tend to occur, (4) how fever affects heart rate or oxygen saturation in COVID-19 patients, and (5) how blood sugars are affected by COVID-19, etc. The addition of environmental and societal variables—such as air pollution level, local temperature, humidity, and population density—adds another layer of complexity and may help us also reveal important epidemiological and environmental factors in COVID-19 transmissibility and phenotype.

Some of the key challenges regarding technological advances in public health has been its deployment and accessibility across populations. Within these two broad areas lie many factors to consider, such as ease of use, cost barriers, platform scalability, etc. One successful case of technology deployed at scale for COVID-19 was NEO, launched by nonprofit organization PhysioQ. It was recognized as one of the world's top COVID-19 innovations by the World Economic Forum [29], and scaled to over 12 countries and 5 languages.

Launched in May 2020, NEO was a free platform (Figs. 1 and 2) that empowered families to early-detect COVID-19 symptoms that simply can't be detected by an individual—such as blood oxygen levels—using affordable consumer wearable devices (i.e. Garmin wristbands/smartwatches).

Through an AI anomaly detection system, families could monitor their vitals together in a private family group and get notified of any group member's health irregularities. In monitoring a variety of biomarkers, such as blood oxygen levels, respiratory rate, heart rate variability (calculated from beat-to-beat R-R interval variations), as well as sleep and activity, a unique baseline was created for each individual. Subtle deviations from their baseline data could potentially indicate that they were

Fig. 1 NEO app and
wearable devices (Garmin)

Fig. 2 NEO web dashboard for family monitoring

coming down with a viral illness [26]. Individuals also had the ability to manually input symptoms, temperature, as well as COVID-19 test results, which served a dual purpose to both flag abnormal results but also to help further train the AI anomaly detection system using a variety of machine learning techniques [30].

In tandem, everyone using NEO had the option to donate their anonymized data to the PhysioQ COVID-19 open Databank, which was made freely accessible to all health researchers worldwide to accelerate research into COVID-19. With 64% of NEO users in the first three months choosing to donate their data, this provided researchers with access to numerous research-quality, real-world physiological datasets [30].

Many initiatives involving wearable device users donating data for COVID-19 research have been extremely successful—from the Corona-Datenspende, developed by the Robert Koch Institut in Germany [31], to the Scripps Research DETECT study [26], to Stanford [32]. However, a majority of the research initiatives globally have kept the collected data within their specific organizations. Unfortunately, this 'siloization' of data significantly limits the speed and breadth of research that could otherwise have been performed by researchers globally had they had access to data [33]. Following PhysioQ's model [30] and increasing the amount of data that is publicly-available for others to analyze can go a long way towards COVID-19-related breakthroughs.

PhysioQ's NEO initiative is an example of how personalized health tools can serve a dual-purpose—to help individuals, but also to empower clinicians, researchers, and public health experts with real-time, real-world data.

Real-time, real-world data for public health experts

We've seen with COVID-19 the importance of quickly making sense of a vast amount of data from various sources, in order to provide prompt and accurate public health responses. Increasingly, public health systems are discovering that the technological tools required to mitigate the toll of the pandemic involve wearables and/or portable technologies. These include home antigen tests, vaccination documentation, contact tracing technologies, and telehealth [34]. Although many still know wearable devices as simple fitness accessories, as successful examples become more widely known, such as PhysioQ's NEO, we should continue to see a shift to incorporating such useful and objective biomarker data into public health strategies. This real-time data collection at scale could then be used by "AI and deep learning systems to understand healthcare trends, model risk associations and predict outcome" [8]. This is a public health game-changer.

In countries where a centralized/integrated health database does not exist, the ability for the public health system to administer these tools, to record and analyze the acquired data and to respond in an expeditious manner is markedly impaired [33].

4 An Equitable Future—Making Personalized Health Tools Accessible to All

With the current COVID-19 pandemic and the growing prevalence of chronic diseases, affordable and effective personalized health tools should be made accessible to both individuals and public health agencies around the world. Although numerous factors are involved in creating health equity and accessibility, we highlight three key elements to enable that future: device-agnostic platforms, data privacy and ownership, and building solutions for LMICs (low- and middle-income countries).

Device-agnostic data collection and AI-analytics platforms

In order for any personalized health technologies, such as wearables, to be deployed by public health agencies at scale, a robust data collection and AI-analytics platform must be utilized. To ensure accessibility to all, such platforms should aim to be device-agnostic, allowing various types of devices at all price ranges to connect.

An example of a device-agnostic platform is nonprofit PhysioQ's LabFront, which allows all types of devices to collect and analyze data for health research. Although launched to democratize access to conducting health research, the same concept can be applied to personalized health solutions.

Data privacy and ownership the key to trust

Data privacy and data ethics are a core element upon which trust is built between any public health agency and the public. It is critical that digital health solutions providers are properly vetted. Oftentimes, despite the fact that the data is generated by the citizens, large corporations are the ones who profit from the data by repackaging the data and selling to other companies. Within this existing paradigm, power resides in these corporations who dictate how the data is to be utilized—and sadly, the primary driver is financial profit. We strongly believe that if this data ownership were shifted back to those who actually generated them—i.e., the patient/citizen, the incentives and motivations would be focused more towards directly benefiting the public—i.e., by advancing general health and well-being.

Accessible to LMICs (low- and middle-income countries)

The use of wristbands/smartwatches (PPG sensors) have been shown useful for the detection and monitoring of various non-communicable diseases (NCD). With the World Health Organization's (WHO) target to reduce deaths from NCD in people age <70 years by 25% by 2025 [35], device-agnostic platforms that allow to integrate low-cost wearable devices (under USD $10) may also offer part of the solution and enable the development of an affordable, sustainable and scalable model of critical care in resource-limited settings [36].

As part of a unified device-agnostic AI-analytic platform, these wearables would also be able to connect with a variety of other tools, such as the Dynamical Biomarkers Groups' 'Tricorder' (XPrize Tricorder finalists) [37]. These types of digital solutions offer remarkable promise to supplement the shortage of human resources in personal healthcare and public health in resource-limited settings.

5 Conclusion

It is argued in Ming et al. [3] that 'Clinical studies have consistently demonstrated the need for timely medical intervention; delayed recognition of illness results in adverse outcomes and increased costs'. This is particularly relevant in this time of the COVID-19 pandemic. With advances in technology over the last decade, from robust cloud infrastructures, to artificial intelligence, to wearable devices, personalized health interventions are becoming increasingly accurate, accessible, and prevalent.

Health management at the individual and public health level in situ and in 'real-time' is now available. This presents significant opportunities to scale up such health monitoring capability to support planning, designing and implementation of public health strategies at the regional, national and global level thereby contributing to sensemaking 'that serves as a springboard into action' [2].

This chapter described the current literature on biomarkers, wearable's and person-alized health. An operational deployment of this capability within the backdrop of COVID-19 was described highlighting both the benefits for personal health and also public health.

References

1. Jackson MC (2019) Critical systems thinking and the management of complexity. John Wiley and Sons, NJ
2. Weick KE, Sutcliffe KM, Obstfeld D (2005) Organizing and the process of sensemaking and organizing. Organ Sci 16(4):409–421
3. Ming DK, Sangkaew S, Chanh HQ, Nhat PTH, Yacoub S, Georgiou P, Homes AH (Jul 2020) Continuous physiological monitoring using wearable technology to inform individual management of infectious diseases, public health and outbreak responses. Int J Infect Dis 96:648–654
4. Evangelatos N, Satyamoorthy K, Brand A (2018) Personalized health in a public health perspective. Int J Public Health 63:433–434. https://doi.org/10.1007/s00038-017-1055-5
5. Soliño-Fernandez D, Ding A, Bayro-Kaiser E et al (2019) (2019) Willingness to adopt wearable devices with behavioral and economic incentives by health insurance wellness programs: results of a US cross-sectional survey with multiple consumer health vignettes. BMC Public Health 19:1649
6. Chen H, Hailey D, Wang N, Yu PA (2014) Review of data quality assessment methods for public health information systems. Int J Environ Res Public Health 11:5170–5207. https://doi.org/10.3390/ijerph110505170

7. World Health Organization. Clinical management of severe acute respiratory infection when novel coronavirus (nCoV) infection is suspected 2020 [Available from: https://www.who.int/publications-detail/clinical-management-of-severe-acute-respiratory-infection-when-novel-coronavirus-(ncov)-infection-is-suspected]

8. Ting DSW, Carin L, Dzau V et al (2020) Digital technology and COVID-19. Nat Med 26:459–461. https://doi.org/10.1038/s41591-020-0824-5

9. Weick KE (1995) Sensemaking in organizations. Sage, Thousand Oaks, CA

10. Klein G, Phillips JK, Rall EL, Peluso DA (2007) A data-frame theory of sensemaking. In: Hoffman RR (ed) Expertise out of context: proceedings of the 6th international conference on naturalistic decision making. pp 113–155

11. Biomarkers Definition Working Group (2001) (2001) Biomarkers and surrogate endpoints: preferred definitions and conceptual framework. Clin Pharmacol Therapeutics 69:89–95

12. BMJ (2015) 350 https://doi.org/10.1136/bmj.h910

13. Dieffenderfer JP, Goodell H, Bent B, Beppler E, Jayakumar R, Yokus M, Jur JS, Bozkurt A, Peden D (June 2015). Wearable wireless sensors for chronic respiratory disease monitoring. In: 2015 IEEE 12th International conference on wearable and implantable body sensor networks (BSN). IEEE. pp 1–6

14. Liu W, Fang X, Chen Q, Li Y, Li T (2018) Reliability analysis of an integrated device of ECG, PPG and pressure pulse wave for cardiovascular disease. Microelectron Reliab 87:183–187

15. Bally L et al (2018) Closed-loop insulin delivery for glycemic control in noncritical care. New Engl J Med 379:547–556

16. Perez MV et al (2019) Large-scale assessment of a smartwatch to identify atrial fibrillation. N Engl J Med 381:1909–1917

17. Thomas RJ et al (2005) An electrocardiogram-based technique to assess cardiopulmonary coupling during sleep. Sleep 28(9):1151–1161

18. Haenssle HA et al (2018) Man against machine: diagnostic performance of a deep learning convolutional neural network for dermoscopic melanoma recognition in comparison to 58 dermatologists. Ann Oncol 29:1836–1842

19. Topol EJ (2019) High-performance medicine: the convergence of human and artificial intelligence. Nat Med 25:44–56

20. Hinton G (2018) Deep learning-a technology with the potential to transform health care. JAMA 320:1101–1102

21. Krittanawong C, Zhang H, Wang Z, Aydar M, Kitai T (2017) Artificial intelligence in precision cardiovascular medicine. J Am Coll Cardiol 69(21):2657–2664

22. Centre for Disease Control and Prevention (2020) Interim clinical guidance for management of patients with confirmed coronavirus disease (COVID-19). [Available from: https://www.cdc.gov/coronavirus/2019-ncov/hcp/clinical-guidance-management-patients.html#foot02]

23. Huang C, Wang Y, Li X, Ren L, Zhao J, Hu Y, Zhang L, Fan G, Xu J, Gu X, Cheng Z (24 Jan 2020) Clinical features of patients infected with 2019 novel coronavirus in Wuhan, China. The Lancet

24. National Institute for Communicable Diseases-South Africa (2020) Clinical management of suspected or confirmed COVID-19 disease. [Available from: https://www.nicd.ac.za/wp-content/uploads/2020/03/Clinical_management_of_suspected_or_acute_COVID_V1.1_13.03.20_updated.pdf]. (https://www.nature.com/articles/s41591-020-1123-x)

25. Oran DP, Topol EJ (2020) Prevalence of asymptomatic SARS-CoV-2 infection. Ann Intern Med https://doi.org/10.7326/M20-3012

26. Quer G, Radin JM, Gadaleta M et al (2021) Wearable sensor data and self-reported symptoms for COVID-19 detection. Nat Med 27:73–77. https://doi.org/10.1038/s41591-020-1123-x

27. Chau NVV et al (2020) The natural history and transmission potential of asymptomatic SARS-CoV-2 infection. Clin Infect Dis https://doi.org/10.1093/cid/ciaa711

28. Kannel WB, McGee D, Gordon T (1976) A general cardiovascular risk profile: the framingham study. Am J Cardiol 38(1):46–51

29. Schwere A (2020) These 15 innovations are helping us fight COVID-19 and its aftermath. World economic forum

30. PhysioQ (22 Dec 2020) An open COVID-19 databank for research, PhysioQ, viewed. <https://physioq.org/databank>
31. Corona Datenspende (Robert Koch Institut) (2020) https://corona-datenspende.de/science/en
32. Mishra T, Wang M, Metwally AA et al (2020) Pre-symptomatic detection of COVID-19 from smartwatch data. Nat Biomed Eng 4:1208–1220. https://doi.org/10.1038/s41551-020-00640-6
33. National Research Council (US) Committee on A Framework for Developing a New Taxonomy of Disease (2011) Toward precision medicine: building a knowledge network for biomedical research and a new taxonomy of disease. National Academies Press; MD, USA
34. Budd J, Miller BS, Manning EM et al (2020) Digital technologies in the public-health response to COVID-19. Nat Med 26:1183–1192. https://doi.org/10.1038/s41591-020-1011-4
35. World Health Organization (2014) Global status report on non-communicable diseases 2014. World Health Organization
36. Turner HC, Hao NV, Yacoub S et al (2019) Achieving affordable critical care in low-income and middle-income countries. BMJ Global Health 4:e001675
37. Center for Dynamical Biomarkers at BIDMC/Harvard Medical School (2017) XPRIZE, DBIOM. <https://dbiom.org/xprize>

Threat Risk Assessment (TRA)
for Physical Security

Darek Baingo

Abstract Maintaining the physical security of an organization entails navigating an intricate landscape of threats, adversaries, systems, and policies. As organizations evolve, become more complex and spatially distributed, security risks increase exponentially and become difficult to fully understand. Organizations entrusted with critical missions and ownership of high risk/high value assets realize that physical protection systems and policies are crucial to prevent unacceptable consequences arising from harmful influences, whether deliberate, accidental or natural. The more serious the consequences, the more important it is to have a high degree of confidence that physical protection will be as effective as planned. The highest level of confidence in physical protection is best achieved through the design and implementation of protective measures that are linked to a thorough understanding of the threats and vulnerabilities. This is achieved through comprehensive and up-to-date analysis of the motivations, intentions, and capabilities of potential adversaries against which protection systems are designed and evaluated. This chapter presents the conceptual development of a Threat Risk Assessment (TRA) Methodology for physical security planning and design. The methodology addresses critical knowledge and capability gaps in TRA approaches, and aims to strengthen the transparency, robustness and defensibility of an organisational Security Risk Management program. The chapter concludes with a discussion of lessons learned and recommendations for future work.

Keywords Physical security · Security design · Threat risk assessment · Vulnerability assessment · Asset prioritization · Design basis threat · Targeting analysis · Risk management

1 Introduction

Detrimental impacts to an organisation may be posed by a very diverse set of risks and the risk environment is never static. New threats emerge, novel means of hostility

D. Baingo (✉)
Centre for Security Science, Defence R&D Canada, Ottawa, Canada
e-mail: darek.baingo@bell.net

© The Author(s), under exclusive license to Springer Nature Switzerland AG 2021
A. J. Masys (ed.), *Sensemaking for Security*, Advanced Sciences and Technologies
for Security Applications, https://doi.org/10.1007/978-3-030-71998-2_14

are being undertaken, climate change has resulted in higher frequency and impacts of natural incidents, and the reliance on critical infrastructure increases rapidly. This Chapter deals with the ideas and concepts behind the development of a new Threat Risk Assessment (TRA) methodology. The aim of this chapter is not to provide all the details of this TRA methodology as that would require more than one chapter. Rather, the objective is to provide an overview, best practice advice and guidance for the development and conduct of comprehensive and scientifically valid TRA methodology for physical security planning and design. The TRA methodology presented here has been informed by reviews of several different TRA methodologies, in-depth discussions with subject-matter-experts (SMEs), security professionals, and reviews of governmental policies. It is hoped that this chapter will inspire others to further explore, discuss and develop comprehensive TRA methodologies.

2 Background

A TRA (often also referred to as a TVRA—Threat Vulnerability Risk Assessment) is essentially a process for identifying and mitigating risk. Many methodologies exist for conducting TRAs and most utilize the following common elements:

- Establishing a **Requirement for the conduct of a TRA** (why is a TRA needed?).
- **Identification of Assets** (e.g., personnel, infrastructure, services, information etc.) and the value of each of those assets to an end objective or mission.
- **Identification and assessment of the threat(s)** for the identified asset(s); for example, from naturally occurring hazards such as earthquakes, floods or a deliberate threat such as the use of an improvised explosive device (IED). The threat assessment often includes an examination of the likelihood of the threat occurring and its impact if it is successful.
- **Assessment of the Vulnerability** of a given asset to the identified threat(s). This includes determining the exploitability or inherent weakness of an asset and to what degree any of the identified threats could impact it.
- **Assessment of Risk** is a consideration of the assets identified, the threat to those assets and their vulnerabilities. In some cases, the vulnerability assessment also includes an assessment of risk.
- The level of **Residual Risk**. This is the remaining risk after solutions to mitigate risks are implemented to a pre-identified acceptable threshold.
- **Recommendations** to mitigate any unacceptable residual risks.

A TRA also identifies what the mission/objective and scope of the TRA will be and what the minimum defined acceptable risk threshold is (typically assigned a descriptor such as low, medium, high etc.).

In general, many organizations are responsible for the conduct of security reviews and TRA's of their assets and facilities. A sustainable TRA program must allow organizations to fulfill their obligations under a Security Plan/Policy. The lack of a standardized TRA methodology and supporting processes can lead to TRAs becoming fragmented across different parts of complex organizations, resulting in security recommendations and physical security design documentation being based on incomplete information and/or poor understanding of how to conduct TRAs using credible scientific methods and best practices. This can lead to expensive or ineffective mitigation measure, or both.

Additional processes, such as security audits, Physical Security Surveys (PSS), Business Continuity Plans (BCP), and Business Impact Analysis (BIA) may also be utilized to inform a TRA. The TRA methodology presented is intended to be used in the following circumstances:

- During capital project definition, development and delivery;
- During system definition, development and delivery;
- For the conduct of a baseline (or initial) TRA; and
- When revisiting a previously completed TRA.

As a starting point, the Harmonized Threat Risk Assessment (HTRA) [1], developed by the Communications Security Establishment (CSE) and the Royal Canadian Mounted Police (RCMP), was used as the baseline methodology and several processes within this baseline were used to guide the development of the TRA methodology presented here. The HTRA has been previously proposed as a standard TRA methodology for use by the Government of Canada. The general processes in the HTRA are:

- Establishing a requirement (to conduct a TRA);
- Scope and definition;
- Asset identification and valuation;
- Threat assessment; and
- Vulnerability assessment

In addition to these processes, the HTRA also explores the following two ideas:

- Identification of various "triggers" that could or should cause a TRA to be conducted or if already completed would cause the reassessment to ensure its continued validity; and
- Discussion of how TRAs should be **prioritized at a strategic level**.

A graphical overview of the processes explored is provided in Fig. 1, and expanded as discussion on each of the processes in the sections that follow.

The information gained during this process was assessed against the results of a comprehensive comparative literature review [2].

2.1 Establishing a Requirement

The decision to carry out a TRA, and the type of TRA to be conducted, needs to be based on established requirements and criteria. There are specialised areas of responsibility that have established defined requirements and protocols for assessing

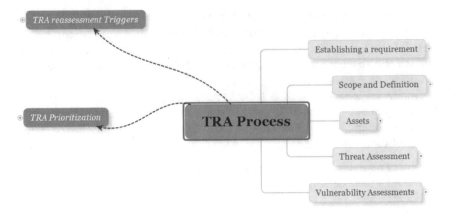

Fig. 1 Overview of TRA process [2]

Fig. 2 Establishing a requirement for TRA conduct

risk and vulnerability. The intent of the discussion during the brainstorming sessions was to identify what aspects of the organisation's operations needed to be included or excluded within the methodology presented. Figure 2 provides a graphical representation of the areas used to guide the discussion.

2.2 Scope and Definition of a TRA

The scope and definition of a TRA is an important consideration. The reason a TRA is being conducted will impact the scope of the assessment and the manner in which it is performed. In some cases, a TRA may be performed to establish a baseline, whereas the scope of another TRA may be based on a change to a single variable. An example of this is where the threat for a previously completed assessment did not include a requirement to analyse vulnerabilities related to a chemical, biological,

Fig. 3 Scope and definition of a TRA [2]

radiological or nuclear (CBRN) attack, but where a change in assessed threat "triggers" a reassessment. The complexity of the TRA will also be impacted by the scope and definition (Fig. 3).

Unless the TRA to be conducted is a baseline assessment, the scope may be closely tied to the trigger for reassessment. The various types of TRAs along with any associated requirement to include them as part of a scalable process in the methodology are described below:

- A TRA conducted for a specific reason or to address a specific threat or concern. The previously sited example of a CBRN threat is a good example. While this may be a formalized process, it may be of reduced complexity and require fewer resources than that a full baseline TRA.
- A Physical Security Survey (PSS) is a type of TRA; however, it is limited in its application for several reasons [2]. Within an organisation the PSS is used to confirm that baseline physical security standards are being applied, which minimizes vulnerability. What the PSS does not do, is determine whether the baseline physical security standards are sufficient in a given situation. For example, the physical security standard may call for video surveillance on all entrances of a building, but may not address additional security requirements based on the nature of a specific threat. Extending the CBRN example, additional cameras with motion sensors and alarms may be needed to monitor ventilation intakes which is above and beyond the baseline security standards.
- Formal or baseline TRAs are those that, as the name implies, provides an initial assessment as to the vulnerabilities, safeguards, and recommended additional safeguards to reduce residual risks to acceptable levels. The baseline TRA methodology should allow for scalability which will allow it to be used for tactical, operational (regional) and strategic assessments. The methodology in question needs to be able to satisfy these requirements.

2.3 Asset Identification

One important issue to be considered is whether both tangible assets (infrastructure, personnel, information etc.) and intangible assets (morale, reputation, psychosocial impacts etc.) should be identified as assets within a methodology, as illustrated in Fig. 2, on the following page.

- The consideration regarding inclusion of the tangible/intangible categories within the identification of assets, is one of whether intangibles such as morale and public confidence are to be considered an asset within the TRA methodology:
- When including intangibles as an impact on an exposed vulnerability, careful attention will need to be paid when incorporating this into the methodology to ensure that there is no "double accounting" of the intangible. That is to say that the consideration of intangible "impacts" is only considered during the vulnerability assessment phase of the TRA;
- Intangible considerations such as effect on morale, impact on public confidence and reputation are by their very definition, subjective in nature. It is important to remove as much of the subjectivity as possible from the equation when considering impact. The use of well thought out word ladders with objective criteria may be one method of reducing subjectivity. This too needs to be considered carefully when incorporating into the methodology (Fig. 4).

In terms of the identification of tangible assets:

- The TRA identifies an organisation's infrastructure critical to the accomplishment of the mission as an asset, while taking into account the interdependency of an organisation's assets on Critical Infrastructure outside the organisation.
- The use/applicability of the TRA methodology under development in an IT context and whether IT should only be considered as an asset for the purposes of this methodology in terms of physical security.
- Tangible assets include infrastructure, equipment, personnel, information and the subsets thereof.

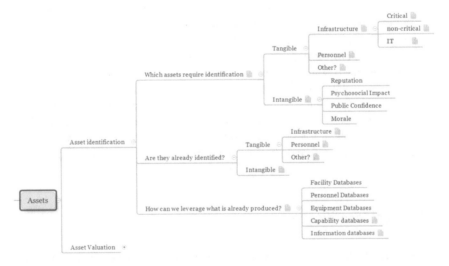

Fig. 4 Asset identification

These processes include capturing of information on assets such as infrastructure, personnel and equipment, the conduct of threat assessments, etc. Additionally, the following considerations need to be reflected upon:

- The aggregation of multiple databases into a single repository that could be exploited by the methodology may introduce additional security risk/vulnerability.
- The BCP/BIA process may provide some measure of support by identifying critical services and hence tangible assets that support those services.

2.4 Asset Valuation

The assigning of values to assets is a critical aspect of conducting a TRA. Asset valuation is based on the impact that would be realized, should the asset be compromised, as shown in Fig. 5. In many methodologies compromise is expressed in terms of unauthorized disclosure (confidentiality), unauthorized removal, destruction or interruption of services (availability) or unauthorized modification (integrity), and finally in terms of monetary impact. The following points of consideration need to be accounted for:

- Asset valuation is routinely thought of as operational importance to the completion of the mission.
- Asset valuation must be standardized across the organisation; removal of subjective assignment of values to the extent possible is important, as assets need to be valuated based on their strategic importance not their locally perceived importance.
- It is beneficial to assign strategic values or categorizations to major infrastructure, equipment etc., based on the organisation's mission(s).
- The valuation of assets requires experience and knowledge. As a result, assigning values to all assets at a strategic level is likely not prudent. Local input into the importance of specific assets needs to be included in the methodology, as local experience and intuition are important aspects of the valuation process.

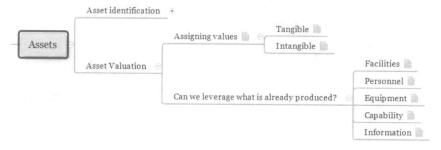

Fig. 5 Asset valuation

- The redundancy of an asset should be considered during the valuation process. A single asset capable of performing a function may have a higher value than if there are many of them capable of fulfilling the same function.

2.5   Threat Assessment

Threat assessments are typically considered in three broad categories. These are *natural hazards*, *accidents* and *deliberate* (malicious/intentional) threats, such as terrorism. When assessing threats, it is accepted practice that the impact of a threat is considered in relation to the likelihood of the threat actually occurring and having an impact (exposure), see Fig. 6. All categories of threat need to be considered in the methodology. Threats are assessed on many levels; locally, regionally, nationally and internationally. They include all facets of threat, such as criminal, kinetic, cyber, homegrown and foreign. Accidental threat is also considered in terms of safety and security; information is collected on accidents from a safety perspective (e.g., slip and falls etc.), as well as on security incidents (loss of classified material, breaches found during a PSS etc.). These are collated for various purposes such as safety analyses and targeting awareness level training. Other considerations include:

- Threat assessments generated for specific TRAs needs to be monitored for any significant change that may warrant the reassessment of a TRA. This could be a new or changing threat, including a change in threat methodology (for example the use of vehicles against crowds).
- Consideration needs to be given to the development and promulgation of a design basis threat with linkages to countermeasures, similar to that used by the Interagency Security Committee (ISC) in the United States [3].
- A TRA specific threat assessment needs to leverage the information from locally produced information such as PSS reports, safety incident reports etc. This may be available at a national level as well.
- There will need to be designated points of contact that will be cleared to receive threat information pertaining to TRAs in their area of responsibility (AOR).

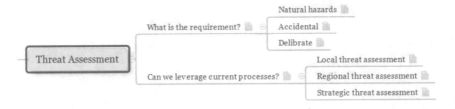

Fig. 6 Threat assessment

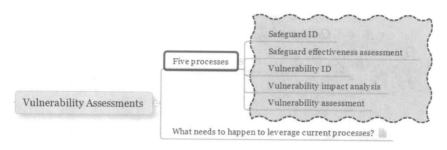

Fig. 7 Vulnerability assessment

2.6 Vulnerability Assessment

To assess vulnerabilities, the assets identified need to be examined in terms of the threat posed to them and the ability of safeguards to protect them. The HTRA uses a five-step process to formulate the assessment (see Fig. 7). The TRA methodology presented herein concentrated on the first three of these processes as they inform the last two processes. Safeguards include such things as physical security standards, IT security, and personnel screening.

During the course of a TRA, asset vulnerability needs to be reassessed whether the threat exposes a vulnerability that physical security standards do not account for. Changes in threat, technology and the assets themselves may increase or decrease safeguard effectiveness. A review of security incidents, results of previous physical security surveys and safety incident databases as part of the threat assessment will help to identify areas that should be concentrated on during TRAs.

2.7 Reassessment Triggers

TRA reassessment triggers are closely related to that of the requirement to conduct TRAs and their scope and definition. The three primary conditions that could trigger a TRA to be revisited are:

1. A change to the threat. This could be an increase or decrease in likelihood, change to attack methodology or technology application, or the emergence of an entirely new threat;
2. A change to an asset in terms of physical construction, or in purpose; and
3. A change in mission.

It should be remembered that a change in one of these areas does not necessarily mean a vulnerability has increased. A change in threat can also mean that the threat has diminished, and a TRA needs to be revisited to ascertain whether the safeguards that were established to reduce vulnerability are still warranted as they may reduce effectiveness and pose an additional burden in terms of convenience or cost. The

Fig. 8 Reassessment triggers

responsibility for determining whether a trigger threshold has been surpassed, and the definition of those thresholds, needs to be included in the policy that support the methodology. This needs to include the reporting network to be followed when trigger thresholds are surpassed (Fig. 8).

2.8 Other Considerations

The following additional considerations are provided to capture issues that do not fall within one of the specific processes:

- The Business Continuity Planning (BCP) process and generation of a Business Impact Analysis (BIA), common corporate tools, may help inform some of the requirements and the scope of TRAs, as they may identify critical services at strategic, and potentially operational and tactical, levels.
- Those participating in the conduct of a TRA need to possess the requisite experience and skill sets, regardless of the methodology being used and that automation of the process may not curtail this requirement, as it is impossible to remove entirely, subjectivity from within the process. While the use of automation, word ladders and other means can be used to help objectify observations and reduce TRA complexity, there will always be a level of intuitiveness involved in the process, and the use of expert opinion, experienced and knowledgeable SMEs is therefore a necessity.
- The scope and definition of a TRA need to be carefully considered, and the methodology being needs to be scalable to allow for use in specific circumstances and at various levels (e.g., tactical, operational and strategic).
- The inclusion of intangibles as impacts rather than assets is preferred.
- Consideration needs to be given to exploiting available current databases for the identification and valuation of assets.
- Asset valuation, to the extent possible, needs to be done objectively and when possible at a strategic level.
- The design basis threat (DBT) needs to be linked to countermeasures, similar to that used by the Interagency Security Committee (ISC) in the United States.

- The TRA methodology needs to use information from other inspections and reporting lines (such as safety), to help identify areas where the baseline physical security standards may not reduce the vulnerability to acceptable levels.
- The responsibility for determining whether a trigger threshold has been surpassed, and the definition of those thresholds, needs to be included in the policy that support the methodology.
- Use of an organisation's BCP process may help inform the TRA methodology in terms of critical services and assets.
- There will always be a requirement to use experienced and knowledgeable individuals to conduct TRAs, regardless of automation.

3 Discussion of Results

The TRA approach presented here was informed by a comprehensive analysis of several established TRA methodologies [2], processes, as well as discussions SMEs. The review of existing TRA methodologies fulfilled two objectives:

1. It provided an overview of an organization's policy requirements to conduct a TRA process; and
2. It identified and described the requirements of a robust TRA methodology.

The comparative review identified and assessed the various TRA methodologies against a set of metrics developed for each of the processes within the HTRA [1] as a common baseline, as summarized in Table 1.

Definitions of these metrics are summarized in Table 2, on the following page.

The other three TRA methodologies investigated but subsequently not selected for inclusion in the comparative analysis are:

- United Kingdom (UK), Center for the Protection of National Infrastructure (CPNI): The CPNI is the UK government authority for protective security advice to national infrastructure. Their role is to enhance national security by helping to reduce the vulnerability of the national infrastructure to terrorism and other threats. The CPNI does not publish/distribute a TRA methodology but describes its features and recommends ISO 31,010 Risk Management—Risk Assessment Techniques [11] as a standard that is available for purchase.
- US DHS—Threat and Hazard Identification and Risk Assessment (THIRA) Methodology [12]: The THIRA is a threat assessment tool intended for use by whole communities. The process involves an analysis of threats/hazards that may occur within a community, along with the risk and consequences of each event. The assessment outcome is a determination of a community's emergency management readiness to respond to crisis and if required, corrective actions/investment. This methodology does not contain relevant processes for inclusion in the TRA methodology presented here.
- The WBDG [13] guide provides government and industry security related guidance, criteria and technology from a 'whole building' perspective. The WBDG

Table 1 Comparison of TRA methodologies based on complexity, adaptability and scalability

Methodology	Complexity	Adaptability	Scalability
RCMP/CSE—Harmonized Threat Risk Assessment (HTRA) methodology [1]	Overly complex	Very adaptable	Very scalable
US DHS/FEMA*—threat, consequence and vulnerability assessment tool [3]	Sufficiently complex	Somewhat adaptable	Somewhat scalable
Ontario—Hazard Identification and Risk Analysis (HIRA) [4]	Not sufficiently complex	Somewhat adaptable	Somewhat scalable
US Interagency Security Committee (ISC) [5–7]	Sufficiently complex	Somewhat adaptable	Somewhat scalable
National Institute of Standards and Technology (NIST) TRA [8]	Not sufficiently complex	Not adaptable	Not scalable
US Department of Justice (DOJ) TRA [9]	Sufficiently complex	Very adaptable	Very scalable
CAF/DND Force Protection Assessment Guide (FPAG) [10]	Overly complex	Somewhat adaptable	Very scalable

*US Department of Homeland Security/Federal Emergency Management Agency

methodology is a whole of building approach. However, it is extremely generic and does not introduce any best practice or methodology relevant to the development of a TRA methodology.

The comparative analysis of the TRA methodologies identified that while there are several commonalities between the various methodologies, there are also distinct processes within the various models which are worthy of consideration. The comparative analysis corresponding to each of the phases in Sect. 2 is presented in more detail in Annex A. This resulted in recommendations and criteria for the development of the final TRA methodology.

3.1 The Proposed TRA Methodology

The TRA methodology incorporates the following sequential phases, as illustrated in Fig. 9 [14]. The first phase of the new TRA methodology is the threat assessment.

Table 2 Subjective metric definitions

Metric	Descriptor	Definition
Complexity	Not sufficiently complex	While easy to use, the lack of detail within the process is likely to result in incomplete or erroneous data opening the process to vulnerabilities
	Sufficiently complex	The process is simple to understand, and of sufficient complexity to likely capture all required information
	Very complex	The process is very intricate and requires automation or adaptation to make useable
	Overly complex	The complexity of the process prohibits its use without significant training, education and frequent usage and automation
Adaptability	Not adaptable	The process is designed for one type of circumstance only (e.g., law enforcement)
	Somewhat adaptable	The process can be used for more than one circumstance but not in all circumstances (e.g., military and law enforcement)
	Very adaptable	The process can be used in most circumstances
Scalability	Perform strategic Targeting Analysis (TA)	The process is specifically targeted at a single asset or type of asset (e.g., IT systems), and/or it is not scalable to different levels (tactical, operational and strategic)
	Determine assets to be prioritized	The process allows some scalability in that it can be applied to a single asset or multiple types of assets (building, IT, personnel) and/or various levels
	Very scalable	The process can be used for all or most types of assets and at all levels

3.1.1 Phase 1—Local Design Basis Threat

The Local Design Basis Threat (LDBT) component of the TRA methodology introduces a unique process for an organisation and establishes an analytical capability to identify and evaluate a full spectrum of threats with the potential to impact organisational operations. The LDBT process supports engagement with

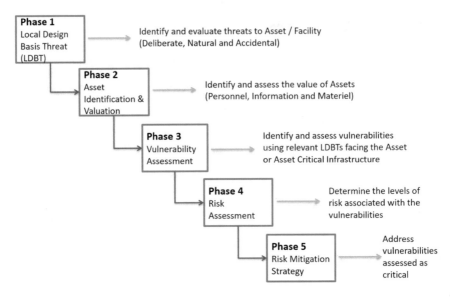

Fig. 9 The phases of the TRA methodology [14]

traditional and new partner agencies in the domestic safety and security community and has been designed to incorporate a scientific approach to threat development [15]. This methodology identifies three categories of threat; Deliberate, Accidental, and Natural. Specific threats within those categories are referred to as **Undesirable Events** (UEs). Examples are vehicle ramming (deliberate), power outage
(accidental), and earthquake (natural).

Incorporation of localised threat into the TRA is important to ensure that the
installation being assessed for vulnerabilities is being assessed against threats relevant
to it. Additional information regarding UEs and incorporation of the LDBT into the
TRA is provided at Annex B.

3.1.2 Phase 2—Asset Identification and Valuation

The Asset Identification and Valuation refers to the development of an inventory of
holdings which an organisation requires in order to perform its core mission(s) and
assigning them a value based on the criticality of those holdings to mission achievement. Holdings are identified in three categories: (i) Personnel, (ii) Information,
and (iii) Materiel and assigns values to each of them based on their importance to
the Asset's ability to accomplish its assigned missions. The values assigned to each
holding category can be used in the calculation of the Asset Statement of Operational
Value ($ASOv$).

3.1.3   Phase 3—Vulnerability Assessment

The vulnerability assessment determines the degree to which the Asset's holdings are vulnerable to identified relevant threats. The process includes a review of unit safeguards including reports and plans related to physical security surveys, accident reports and Standard Operating Procedures (SOPs). The vulnerability assessment documents and assigns Asset Vulnerability values (AVv) which are used to calculate the Asset Risk rating (ARr).

The Vulnerability Assessment process will:

- Identify Undesirable Events (UE) applicable to the Asset;
- Determine which UEs are applicable to specific infrastructure;
- Calculate the Asset Threat value (ATv);
- Assess the physical safeguards, plans, procedures, etc., that are in place to counter the identified threats based on the Vulnerability Assessment Plan;
- Establish the Unit Vulnerability value (AVv) that are applied during the calculation of Risk.

3.1.4   Phase 3—Risk Assessment

The risk assessment phase of the TRA methodology determines the levels of Risk associated with the vulnerabilities identified during the vulnerability assessment phase [14]. The risk assessment conducted during this phase of the TRA is consistent with the organisation's security directives, which generally stipulate that security risk assessments will identify the significance of risk, determine whether the risk is acceptable or not and ascertain if it requires treatment.

The formula for calculating the Asset Risk is:

$$ARr = ATv \times ASOv \times AVv$$

where

ARr Asset Risk rating;
ATv Asset Threat value;
$ASOv$ Asset Statement of Operational value; and.
AVv Asset Vulnerability value.

Each of the three values (ATv, $ASOv$, AVv) used in the unit risk calculation are obtained from phases 1 through 3 respectively. Once the Asset Risk rating is calculated, the numerical value is cross referenced to a risk level spectrum to determine the Risk Level.

3.1.5   Phase 3—Risk Mitigation Strategy

The final phase of the TRA Methodology is to develop a risk mitigation strategy aimed at reducing the Asset Risk rating to lower levels and ensuring that any critical safeguard deficiencies are addressed. Recommendations may include improving basic physical security safeguards, altering plans or procedures, or incorporating enhanced security safeguards. The risk mitigation strategy includes the provision of recommendations for strengthening safeguards, decreasing vulnerability and mitigating risk to a standard commensurate with the operational protection requirements of the Unit being assessed. The Initial Risk rating is the risk score that is attributed to the Unit being assessed until the mitigation recommendations have been implemented; the value is then recalculated using the new Unit and Infrastructure vulnerability values, resulting in the Residual Risk rating. The identified critical safeguard deficiencies and resultant risk mitigation recommendations are the foundation of the final TRA report.

3.2   Selection and Prioritization of Assets for Conduct of a Baseline TRA

Central to the conduct of a TRA is the requirement to develop a process for the selection and prioritization of assets [14]. While not part of the TRA methodology as such, this essential stand-alone process has been designed to establish the priority for the conduct of a baseline TRA. For a single Asset organisation this step is not necessary as the prioritization is obvious. However, for organisations with many Assets (whether in a shared campus environment) or geographically distributed nationally or internationally, the prioritization of Assets is critical. The selection and prioritization process are unique attributes of this TRA methodology and are based on several overarching strategic principles, including transparency, defensibility and repeatability (to the greatest extent possible), and can also be used to justify the prioritization of security related expenditures, utilizing exigent information to the degree possible [15]. The asset prioritization criteria should be directly linked to the mission(s) of the organisation. The selection and prioritization process consists of six steps, as summarized in Table 3.

The selection and prioritization process summarized in Table 3 is used to evaluate all assets. Steps 1 and 2 are achieved through working groups (WGs) with senior management and SMEs, resulting in the strategic valuation of the organisation's assets.

The Local Design Basis Threat (LDBT) leverages intelligence, law enforcement and scientific information to develop local threat profiles relevant to the asset (Step 3), and the Strategic Targeting Analysis (TA) is applied to rate the susceptibility of the facilities/assets to deliberate attack, exposure to natural hazards and accidental events (Step 4). The detailed procedures for developing the LDBT can be found in

Table 3 Selection and prioritization process for baseline TRA conduct and output summary

Step	Purpose	Key output
1	Update the Asset Categorisation List (ACL)	Up-to-date ACL and list of sites to be prioritized
2	Perform Asset Strategic Valuation	Strategic valuation of all assets (ranking)
3	Incorporate the Local Design Basis Threat (LDBT)	Threat assessment value for each asset, reflecting local threat analysis
4	Perform Strategic Targeting Analysis (TA)	Targeting analysis score for each asset
5	Determine assets to be prioritized	Prioritized list of assets for baseline TRA
6	Calculation of Prioritization Score (PS)	Completed prioritization list

[16]. The threats and hazards are related to an assessment of evolving and emerging Undesirable Events (UE)—Annex B. Incorporation of the localised threat into the TRA is performed by utilising the individual UE threat ratings, developed as part of the LDBT methodology. Use of the LDBT is essential to ensure that the asset being assessed for vulnerabilities is being assessed against threats relevant to it.

In order to perform the Strategic Targeting Analysis (TA) a hybrid model fusing the US military MSHARPP (Mission, Symbolism, History, Accessibility, Recuperability, Population and Proximity) and CARVER [17] (Criticality, Accessibility, Recuperability, Vulnerability, Effect, and Recognisability) methods was developed. Both CARVER and MSHARPP were originally developed by the US military for targeting analysis. The MSHARPP model looks at targeting from the perspective of the infrastructure owner only and not from that of the adversary while the CARVER approach examines targets from the perspective of the attacker. Steps 5 and 6 are performed by SMEs to develop the prioritized list of facilities installations for the conduct of baseline TRAs. The hybrid TA model summarized in Annex C is based on maximizing the benefits and minimizing the weaknesses of each model.

4 Recommendations

It is recommended that a list of enhanced physical security safeguards (i.e., standards that exceed the baseline physical security standards already in place) be developed and maintained as a living document. It is thought that the introduction of such a list could help to ensure commonality in approach to resolving unacceptable risk, reduce redundancy in effort and could include a cost/benefit analysis especially when more than one solution exists is of significant importance to decision-makers. In addition:

- A single entity within the organisation should be responsible for receiving and collating TRAs for each asset to identify common security risk management related issues at a strategic level and for input in an organisation's "Risk Register".

- The methodology developed may or may not need to include information technology (IT) system development and maintenance processes but it is applicable to physical security of IT systems.
- Regardless of the methodology employed, the use of experienced, capable and well-trained personnel to conduct the TRA is essential, as subjectivity cannot be entirely removed from the process. That being said, there is a requirement to develop and institute appropriate training for staff designated to carry out the TRA process.
- The Business Continuity Planning (BCP) process, and more specifically the Business Impact Analysis (BIA), should be structured and conducted in such a way that the information they produce, such as the identification of critical services, is easily adopted into the TRA methodology.
- It is suggested that intangibles such as morale, public confidence, and psychosocial impact no longer be considered as assets, but rather that they be considered in terms of impact, resultant of a compromise to a tangible asset.

5 Conclusion

The TRA methodology presented in this chapter was developed to specifically examine Assets (personnel, buildings, critical infrastructure) owned by the organisation performing the assessment in a manner that assesses the impact of security threats on operations/organisational mission. The inclusion and use of properly interpreted intelligence/crime information is an essential element of an organisation's mission profiles (specific threats and hazards). The TRA methodology presented incorporates this information to ensure the rigour and accuracy of physical security TRAs. The methodology provides a specific step-by-step approach to TRA conduct, thereby reducing the requirement for extensive training. It is used to identify and utilize data sets that are already in use, thus reducing duplication of effort.

Although the proposed TRA methodology can be implemented in a "spreadsheet" type of application, it is ideally suited for an automated computational environment. Effort should be invested in developing effective visualization of the results of TRAs to best support decision-makers. Finally, this TRA methodology can pave the way for future collaboration with the larger community of organisations including owner/operators of critical infrastructure and government entities (such as Public Safety, Security, and Emergency/Disaster risk management).

Annex A: Comparative Analysis of TRA Methodologies

This Annex contains the comparative descriptions of TRA methodologies as related to each phase of the baseline HTRA using the metrics identified in Sect. 2.

Requirement and Scope Definition Phases

Requirements

There was a wide variance in how assets were identified and classified/categorized between the various methodologies. However, the requirement to conduct a TRA falls out of policy direction, which should be based on specific criteria, such as:

- large capital project definition, development and delivery;
- system development; and
- establishment of an organisation's baseline.

The establishment of a baseline TRA for critical infrastructure, equipment, personnel, information etc., is widely recognized as a best practice. Once a baseline TRA has been completed, the requirement to revisit or revalidate a TRA may be caused by any number of "triggers". Roles and responsibilities to identify and act upon these triggers also need to be instituted. Triggers include but are not limited to:

- change in any aspect of threat;
- changes in asset inventory or purpose;
- identification of safeguard failures; and
- a cyclical review process.

Whether the TRA model under development will be used for IT system implementation and maintenance is unclear and requires resolution. What is clear is that all TRA models in use need to use the same terminology and key definitions, and reporting should be funnelled through a central national level OPI via the appropriate chain of command.

Scope and Definition

Scalability of a TRA methodology was deemed to be of critical importance. It is clear that a TRA may be required for various reasons, and at various levels within the organisation. This means that a TRA may be required for a single asset (e.g., high value stand-alone building), at a regional/international level or strategically. Similar to the aforementioned need to clarify the organisational policy regarding the requirement to conduct a security risk assessment, clear policy direction regarding the responsibility for determining the scope and definition is also required. It is important that local staff be engaged for their input during this phase, as incorporation of local knowledge during this phase will be valuable in ensuring that the correct assets are being assessed for the right reasons.

Asset Identification and Valuation Phase

Asset Identification

During the brainstorming sessions, the question as to whether to include both tangibles and intangibles as assets in the new methodology was debated at length. Among the models reviewed, only the HTRA identified intangibles (e.g., morale, public confidence etc.) as assets. It was decided that intangibles would be assessed as a consequence or impact, rather than an asset, as part of the new methodology.

There are numerous processes for the identification of assets, and at an even more basic level, how to define an asset. For those models that did identify (tangible) assets, they were usually defined in terms of the following categories:

- Personnel;
- Information;
- Facilities;
- Infrastructure (e.g., equipment/systems); and
- Services.

During the brainstorming sessions, it was determined that for the Departmental methodology, services or capabilities would not be considered as a category of tangible asset; the first four categories are required to provide services, and hence the compromise of these would result in an impact on the ability to deliver of a service.

Several of the methodologies provided lists of assets which served as an aide memoire, providing examples of the various types of assets within each of the categories. Within the department there are already databases of some categories of assets (e.g., personnel, facilities, equipment etc.), that may be exploitable for the purposes of identifying them within a specific geographical location (i.e., base/wing). There was some discussion during the brainstorming sessions that perhaps valuation of an asset in terms of operational criticality should take place first, allowing for identification and selection of critical assets during the scope and definition phase. Using the ISC model's Facility Security Level (FSL) [6] as a baseline, the development of a tool for determining and assigning a facility's (or asset's) value, is worthy of consideration.

Asset Valuation

There were two main focuses of the various methodologies reviewed in terms of asset valuation. One approach was to define the criticality of the asset using various language ladders or parameters, while the other approach focussed on assigning a value to an asset by determining the impact of its compromise using parameters such as confidentiality, availability and integrity. Both these methods involved some degree of subjectivity. The latter process was much more complex, albeit more granular.

It was stressed during the brainstorming sessions that the removal of as much subjectivity as possible from the asset valuation process was very important. This would indicate that the development of a strategic baseline asset valuation table (which would also require the development of a strategic asset identification table) would be a useful tool, similar to the FSL developed as part of the ISC documentation. However, assigning values to all assets at a strategic level is likely not prudent. Local input into the importance of specific assets needs to be included in the methodology, as local experience and intuition are important aspects of the valuation process. Therefore, a combination of both approaches would likely be of the greatest benefit. Another important consideration in terms of asset valuation is asset redundancy.

Similar to asset identification, the use of current databases for extracting asset values, at least in monetary terms and, likely in other terms as well, may be feasible and should be investigated. Asset valuation may be tied to the BCP process, and more specifically the BIA which identifies critical services albeit from a corporate perspective.

Threat Assessment Phase

Threat assessments were for the most part considered in three broad categories. These are natural hazards, accidents and deliberate threats such as terrorism. When assessing a threat, it is accepted practice that the impact of a threat is considered in relation to the likelihood of the threat actually occurring. Use of word ladders seemed to be an effective method of reducing subjectivity.

Within the Department, the threat picture is continuously monitored, and the products developed (or at least the process used to develop them) should be leveraged for use in TRAs. For this current capability to be leveraged, TRA threat data requirements will need to be defined. The development and use of a DBT product are worthy of investigation, and in fact may be a best practice. This could include the development of a baseline TRA threat assessment template for completion by intelligence circles; collection and inclusion of information from other federal departments on natural hazards such as earthquakes etc., needs to be explored further. For the production of this type of threat product to be generated, roles and responsibilities will need to be assigned both within intelligence circles and for those that would need to be in receipt of these products. Additionally, a "trigger" mechanism to alert that a change in threat (i.e., probability, modus operandi etc.) may require a TRA to be revisited, should be considered for inclusion in the methodology of the future. This would necessitate that the process allows for the variation of threat over time.

Once the baseline threat is developed, it would need to be reviewed and adapted for use at the local level. Other local threat metrics are already produced and should be exploited. Examples include safety reports, security violation reports and PSS reports; the information provided in these types of products should be considered when developing the local threat picture. The local threat picture should also include crime statistics and trends.

The review process indicated that an increased complexity in the threat assessment process resulted in greater scalability and adaptability; careful consideration needs to be given to the scalability and adaptability requirements of the methodology under consideration. Given the confined context of the intended application (DND/CAF versus all of government), a reduction in scalability and adaptability may be entirely warranted, which may then result in a corresponding reduction in the complexity required for the development of threat assessments.

Vulnerability Assessment Phase

All of the methodologies that used a vulnerability assessment phase, used on-site assessments and interviews as a major data gathering process. Assessment of the data ranged from very complex to being of insufficient complexity. Once again complexity could be correlated to what the methodologies were trying to address in terms of assets, and the impact on scalability and adaptability was similar to that of the threat assessment process; a reduction in complexity may be warranted by the reduced requirement for adaptability and scalability in the global sense.

All methodologies used some type of language ladder to define, and in some cases quantify, impacts/vulnerabilities. It is suggested that an adaptation of this process would also work best for the methodology in question.

During the brainstorming sessions, it was clear that there was no requirement to reassess minimum security safeguards; this was captured quite adequately by the PSS. There is a requirement however, to assess whether the minimum safeguards are effective in countering the current threat, thus reducing vulnerability to an acceptable level.

Reviewing PSS results, safety and security reports etc., prior to the conduct of the TRA, may assist in exposing possible vulnerabilities, and provides some measure of the ability of baseline security/safety measures address the threat.

Risk Assessment Phase (and Calculation of Residual Risk)

While these two phases were presented as distinct processes, they both essentially generate the same information; a descriptor of what the risk is after safeguard effectiveness has been factored in. The formulas used are entirely dependent on the processes that have preceded this phase within each methodology. The following three formulas are provided for comparison purposes:

1. FEMA—the risk assessment integrates the likelihood/probability of the attack (threat) occurring with the probability that a successful attack will produce consequences of a certain magnitude, given the vulnerabilities of the target:

$$\text{Risk} = \text{Threat} \times \text{Vulnerability} \times \textbf{Consequences}$$

2. HTRA—this tool uses values for frequency, impact/consequences and a "change in risk" factor (a correction based on predicted changes to frequency and vulnerability).

$$\text{Risk} = \text{Threat} \times \text{Vulnerability} \times \textbf{Asset Value}$$

3. The DOJ risk is calculated by the following equation:

$$\text{Risk} = \text{Threat} \times \text{Vulnerability} \times \textbf{Criticality}$$

For Residual risk the HTRA assigns numerical scores from 1 to 5 to asset value, threat and vulnerability, based on their assessed descriptor score of very low to very high. Residual Risk is the product of Asset × Threat × Vulnerability. In both cases it was assessed that these calculations provide a sufficient scale on which to base mitigation decisions. The complexity of the calculation was assessed as very complex.

Risk Mitigation Strategy (Recommendations Phase)

The identification of unacceptable risks and the provision of recommendations to mitigate those risks to acceptable levels, are the critical outputs of the TRA. The method used to determine what constitutes an unacceptable risk varied from a consultative process to the use of a Target Risk Level (e.g., medium) as the acceptable level of risk, hence anything above medium is unacceptable.

The identification of additional safeguards to reduce risk, requires that the personnel used for this process have an in depth and up-to-date understanding of what the possible remedies may include. Additionally, the inclusion of a cost benefit analysis, especially when more than one solution exists, is of significant importance for decision makers.

The ISC use of Facility Security Levels (which could be looked at as facility criticality levels), uses weighted values for various parameters such as facility size, importance, population, replacement value etc., to establish a priority rating for buildings; recommendations from the TRA are assessed against the FSL and additional documentation that lists physical standards for each FSL/residual risk situation.

The use of TRLs, and the incorporation of a process similar to the use of FSLs, as well as the development of a list of corresponding additional safeguards to reduce risk levels to acceptable standards, based on the criticality of the asset, is worthy of further investigation.

Annex B: Localized Design Basis Threat

Overview of the LDBT and Use in the TRA Methodology

Undesirable Events (UE)

The DBT identifies and rates "Undesirable Events" (UEs) based on their likelihood and impact. A UE will fall into one of three categories of threat types; **Deliberate**, **Accidental** or **Natural**, with each UE identifying and defining a specific threat event. Some examples of UE are provided in Table 4 (Deliberate UE), Table 5 (Accidental UE) and Table 6 (Natural UE).

The LDBT represents the UE that are considered relevant to the Assets being assessed for a TRA. The LDBT is based on the Regional DBT which is assessed locally to ensure that the local threat environment is accurately represented. UEs identified in the Regional DBT are evaluated by the TRA Team Leader and appropriate Asset representatives; the resulting UE ratings are applied to the TRA process.

Table 4 Examples of deliberate undesirable events

Deliberate	

Ballistic attack	• Active intruder
	• Standoff weapons attack
	  – RPG
	  – Sniper
	  – Mortar
	  – UAV/Drone use for weapons delivery

Vehicle ramming	• Aircraft as a weapon • Automobile ramming
Explosive/incendiary device	• Arson/fire as a weapon • Person borne IED—internal • Person borne IED—external • Parcel IED (mail or delivery) • VBIED
Criminal activity	• Civil disturbance • Theft of weapons or explosives • Theft of classified material
CBR attack	• External • Internal • Mail or delivery • Water supply

Table 5 Examples of accidental undesirable events

Accidental	
Transportation accident	• Train derailment with release of toxic chemicals • Aircraft crash
Extended power outage	• Post storm • Human error
Building/structural collapse	• Resultant from Seismic activity, building misuse • Lack of lifecycle management—deterioration
Hazardous materials incident	• Refinery explosion or other release • Truck accident with HAZMAT release
Nuclear emergency	• Nuclear powered or capable vessel accident • Commercial power/research reactor accident

Table 6 Examples of natural undesirable events

Natural	
Extreme weather	• Geomagnetic storm • Hurricane • Tornado(s) • Blizzard • Ice storm
Flood	• Prolonged rain • Rapid thaw of significant snow
Earthquake	• Of sufficient magnitude to cause structural damage and/or injuries
Landslide	• Heavy rains/deforestation/post wildfire
Forest/wild land fire	• Of sufficient size and proximity to be considered a threat

UE Ratings

Each UE is assessed regionally to determine its relevance to the Unit being assessed and subsequently "localised" to ensure that the UE ratings being used for the TRA represent local influences. As an example, the UE for "Earthquake" may have a higher probability of occurring (and severity/impact) in Esquimalt than Halifax. As a result, this UE may be rated "High" in Esquimalt and "Low" in Halifax. Table 7 provides the values and associated ratings for individual UEs, while Table 8 provides normalised Unit threat values and their associated ratings.

An accurate DBT assessment is critical to the success and credibility of the TRA process. In this methodology, the DBT provides the threat values necessary inform

Table 7 Example individual
UE rating table

Value	UE rating
4.5–5	Very high
3.5–4	High
2.5–3	Medium
1.5–2	Low
Less than 1.5	Very low

Table 8 Example
normalized UE rating table

Value	UE rating
8.9–10.00	Very high
6.60–8.89	High
4.40–6.59	Medium
2.10–4.39	Low
Less than 2.1	Very low

the vulnerability assessment and calculate risk. The TRA team leader should be well versed with the DBT methodology

Annex C: Targeting Analysis

The targeting analysis process uses six Target Selection factors (TSFs) that were developed specifically for the TRA methodology. Care was taken to ensure that TSFs assessed did not duplicate criteria otherwise accounted for in the NDBT and Strategic Asset Type valuation processes

Specific rating criteria have been developed for each TSF. Each rating criteria represents a specific variable associated with susceptibility factors relevant to a specific TSF. Each rating criteria within a TSF contains language ladders which have been designed to remove as much subjectivity as possible; the most accurate descriptor for each rating criteria is selected which will result in a value being assigned. The TSFs developed for use in the prioritization methodology are presented below

- TSF 1—**Symbolism**: Assesses the degree to which an installation symbolises—or may be perceived by an aggressor to symbolise—the Government of Canada (GoC), the Canadian military, Canadian interactions with foreign militaries, or other activities associated with the installation which may represent symbolic targets to an attacker.
- TSF 2—**Recognisability**: Assesses the degree to which an installation's local and/or regional profile will impact its susceptibility to a UE. An installation's

profile includes its physical footprint, the degree to which it is a visible presence in the community where it is located, and its identifiable attributes.

- TSF 3—**Accessibility**: Unfettered access to an installation and the ability to conduct undetected pre-attack surveillance on it increase its susceptibility to a UE. TSF 3 assesses the ease with which an installation can be accessed by examining the daily access control posture of the installation and identifying factors related to the ease with which surveillance could be conducted. It also examines the installation's compliance with Departmental Security Policy related to physical security.
- TSF 4—**Proximity**: Examines the susceptibility to a UE related to an installation's surroundings, including identifying factors associated with the neighbourhood in which the installation is located. Population density can be associated with the amount of pedestrian and vehicle traffic in the vicinity of an installation, increasing potential victims of a UE and/or potential concealment of an attacker. The identification of other potential nearby targets in the vicinity of the installation increases the attractiveness of an attack in the vicinity and accordingly on the installation.
- TSF 5—**Recuperability**: An installation's susceptibility to a UE is considered to increase as its ability to recover or recuperate from it decreases. TSF 5 assesses the installation's ability to maintain its core mission by identifying its ability to mount response operations which can mitigate damages and/or injury and by identifying the scale of redundancies available to it, which are necessary to resume regular operations associated with mission accomplishment.
- TSF 6—**Demographic**: Examines the susceptibility of an installation to a UE based on its demographic, including an assessment of military personnel composition, civilian employee composition, visitation/access to members of the general public, and the visit history of foreign military members, dignitaries, or Internationally Protected Persons (IPP).

All 6 TSF ratings for each installation are added together to calculate the Asset's Targeting Analysis value (ATv), which is used in the calculation of the Prioritization Score. A higher ATv represents a higher assessed susceptibility to a UE.

References

1. Harmonized Threat Risk Assessment Methodology (2007) Ottawa, Royal Canadian Mounted Police, Communications Security Establishment (TRA-1), Available at: https://cyber.gc.ca/en/guidance/harmonized-tra-methodology-tra-1. Date accessed on 22 Jan 2021
2. Morris R, Bureaux J, McDonald S (2017) A comparative review of the threat risk assessment methodologies, defence research and development Canada, DRDC-RDDC-2017-C130
3. Facility security plan: an interagency security committee guide, 1st edn, US, 2015-02, Interagency Security Committee (ISC)
4. Reference manual to mitigate potential terrorist attacks against buildings, 2nd edn, 2011-10, Department of Homeland Security.

5. Hazard identification and risk assessment workbook, Toronto, Ontario, 2012, Emergency Management Ontario
6. The design-basis threat, an interagency security committee report, 9th edn, US, 2014-08, Interagency Security Committee (ISC)
7. Risk management process of federal buildings, 2nd edn, 2016-11, Interagency Security Committee (ISC)
8. Guide for conducting risk assessments, Gaithersburg, Maryland, 2012 Sept, US Department of Commerce, National Institute of Standards and Technology
9. Assessing and managing the terrorism threat, Washington, DC, 2005, US Department of Justice
10. Canadian Armed Forces (2007) B-GJ-005-314/FP-050 CF force protection assessment guide (FPAG). National Defence, Ottawa
11. ISO 31010:2019 Risk management–risk assessment techniques, International Organisation for Standardisation, 2019
12. Threat and hazard identification and risk assessment guide—comprehensive preparedness guide, 2nd edn, Aug 2013, Department of Homeland Security
13. National institute of building sciences, whole building design guide (WBDG) methodology: https://www.wbdg.org/. Date accessed on 20 Jan 2021
14. Baingo D, Friesen SP (2020) Threat risk assessment implementation project, defence research and development Canada, DRDC-RDDC-2020-L226
15. Morris R, Bureaux J, McDonald S (2018) Draft methodology for the development of a DND/CAF national design basis threat assessment, Defence Research and Development Canada, DRDC-RDDC-2018-C238, Dec 2018
16. Morris R, Bureaux J, McDonald S (2018) Procedures for the selection and prioritization of DND/CAF installations for baseline TRA conduct, Defence Research and Development Canada, DRDC-RDDC-2018-C273, Dec 2018
17. Field Manual 3-37.2 Antiterrorism, headquarters department of the Army, Feb 2011. www.us.army.mil. Accessed on 12 Oct 2020

Crisis Leadership and Sensemaking

Anthony J. Masys

Abstract Events such as 9/11, Fukushima Daiichi Nuclear accident, Deepwater Horizon Oil disaster, Hurricane Katrina, Ebola outbreak in West Africa, 2003 US/Canada Blackout, global financial crisis of 2008 and the COVID-19 pandemic all contain the hallmarks of a crisis event. Weick and Sutcliffe (Managing the unexpected: resilient performance in an age of uncertainty. Wiley, San Francisco, 2007 [1]) argue that 'unexpected events often audit our resilience. Everything that was left unprepared becomes a complex problem'. Leadership plays a pivotal role in managing such crisis (whether they emerge as a black swan event (Taleb in The black swan: the impact of the highly improbable. Penguin Books Ltd, London, 2007 [2]), the 'elephant in the room' or as a 'creeping crisis' (Boin et al. in Risk Hazards Crisis Public Policy 11(2), 2020c [3]). The Covid-19 pandemic for example has resulted in considerable impact on global health security, human security and the global economy. As described in (OECD in A systemic resilience approach to dealing with Covid-19 and future shocks New Approaches to Economic Challenges, 2020 [4]), '…the pandemic has reminded us bluntly of the fragility of some of our most basic human-made systems. Shortages of masks, tests, ventilators and other essential items have left frontline workers and the general population dangerously exposed to the disease itself. At a wider level, we have witnessed the cascading collapse of entire production, financial, and transportation systems, due to a vicious combination of supply and demand shocks'. This highlights the volatility, uncertainty, complexity and ambiguity (VUCA) conditions that shape the crisis landscape and its effect on inherent vulnerabilities that exist within our systems. By drawing upon recent disaster events and more recently the COVID-19 pandemic, this chapter examines sensemaking within the context of crisis leadership and presents a crisis leadership framework focused on absorptive, adaptive and generative capacities (Castillo and Trinh in J Organ Change Manage 32(3), 2019 [5]) to support problem framing, solution navigation and innovation.

A. J. Masys (✉)
College of Public Health, University of South Florida, Tampa, FL, USA

International Centre for Policing and Security, University of South Wales, Pontypridd, Wales, UK

© The Author(s), under exclusive license to Springer Nature Switzerland AG 2021
A. J. Masys (ed.), *Sensemaking for Security*, Advanced Sciences and Technologies for Security Applications, https://doi.org/10.1007/978-3-030-71998-2_15

Keywords Leadership · Crisis · Disaster · Complexity · Sensemaking · HRO · Generative leadership

1 Introduction

Events such as 9/11, Fukushima Daiichi Nuclear accident, Deepwater Horizon Oil disaster, Hurricane Katrina, Ebola outbreak in West Africa, global financial crisis of 2008 and the COVID-19 pandemic all represent the hallmarks of a crisis event. The aetiology of these events has been discussed in Masys et al. [6], Masys [7–13], Masys et al. [14] and provides the backdrop for emergent themes regarding leadership in crisis situations. Leadership plays a pivotal role in managing such crisis (whether they emerge as a black swan event [2], the 'elephant in the room' or as a 'creeping crisis' [3]. The Covid-19 pandemic for example has resulted in considerable impact on global health security, human security and global economy. As described in OECD [4, p. 2] 'the pandemic has reminded us bluntly of the fragility of some of our most basic human-made systems. Shortages of masks, tests, ventilators and other essential items have left frontline workers and the general population dangerously exposed to the disease itself. At a wider level, we have witnessed the cascading collapse of entire production, financial, and transportation systems, due to a vicious combination of supply and demand shocks'. Boin et al. [3] refers to this as a 'creeping crisis'. Boin et al. [3, p. 119] argues that 'Practitioners often find it hard to recognize the devastating potential of these creeping crises. In some cases, they do not seem to recognize the creeping crisis at all (manifestation of the creeping crisis is met with shocked surprise). In other cases, practitioners address what in hindsight turn out to be mere symptoms, allowing the bubbling crisis to develop its devastating potential under the radar'. Recognition and understanding of the crisis dynamics is central to sensemaking and crisis leadership.

This certainly connects with Weick and Sutcliffe [1] reflection: '…unexpected events often audit our resilience. Everything that was left unprepared becomes a complex problem'. COVID-19 is stress testing our national and global societal systems and revealing the inherent vulnerabilities that reside within these systems. Taleb [15] argues that 'not seeing a tsunami or an economic event coming is excusable, building something fragile to them is not'. The COVID-19 outbreak along with other public health crisis over the last few years have unveiled major deficiencies in preparedness, response and recovery initiatives locally, regionally and globally. This points to the importance of sensemaking across the COVID-19 pandemic disaster/crisis management cycle (Mitigation, Preparedness, Response, Recovery).

The threat and risk landscape that shapes crisis events emerge and exist in a world characterized by Volatility, Uncertainty, Complexity and Ambiguity (VUCA) conditions. With global societal systems being more interconnected and interdependent, it becomes increasingly difficult to identify causes and effects of complex problems [10, 11, 16, 17]. This presents challenges to problem framing and solving for leadership under crisis conditions. Boin et al., [18] defines crisis management

'...as the sum of activities aimed at minimizing the impact of a crisis. Impact is measured in terms of damage to people, critical infrastructure, and public institutions. Effective crisis management saves lives, protects infrastructure, and restores trust in public institutions'. Effective leadership is thereby a critical requirement in managing crisis. Canyon [19, p. 8] describes crisis leadership as, 'The capacity of an individual to recognize uncertain situations that possess latent risks and opportunities to ensure systematic preparedness, to discern necessary direction, to make critical decisions, to influence followers and to successfully eliminate or reduce the negative impact while taking full advantage of positive aspects within a given timeframe'. The current COVID-19 pandemic has certainly stress-tested our crisis management capabilities. Resonating with the definitions above, lessons learned pertaining to crisis management and COVID-19 are concurrently being captured, providing insights into essential characteristics and leadership qualities required [20].

Threat, uncertainty, and urgency certainly capture the characteristics of a crisis [21, p. 2]. Leveraging the insights from Canyon [19, p. 8], crisis leadership can be decomposed into 4 main characteristics: **sensemaking** (recognize uncertain situations); **systems thinking** (understanding latent risks, impact analysis); **operationalization** (decision making, optimization); **influence and engagement** (working with teams and stakeholders, foster innovation). With this in mind, this chapter presents a crisis leadership framework focused on absorptive, adaptive and generative capacities [5] to support problem framing, solution navigation and innovation.

2 Crisis Defined

The term crisis is defined throughout the literature, rooted in a particular lens or application. As described in Boin et al. [21, p. 1]:

> 'Crises come in many shapes and forms. Conflicts, man-made accidents, and natural disasters chronically shatter the peace and order of societies....It refers to an undesirable and unexpected situation: when we talk about crisis, we usually mean that something bad is to befall a person, group, organization, culture, society, or, when we think really big, the world at large. Something must be done, urgently, to make sure that this threat will not materialize. In academic discourse, a crisis marks a phase of disorder in the seemingly normal development of a system.... Our definition of crisis reflects its subjective nature as a construed threat: we speak of a crisis when policy makers experience "a serious threat to the basic structures or the fundamental values and norms of a system, which under time pressure and highly uncertain circumstances necessitates making vital decisions'.

Military Context

> Crisis—An incident or situation involving a threat to the United States, its citizens, military forces, or vital interests that develops rapidly and creates a condition of such diplomatic, economic, or military importance that commitment of military forces and resources is contemplated to achieve national objectives. (JP 3-0) [22]

UN WHO

A situation that is perceived as difficult. Its greatest value is that it implies the possibility of an insidious process that cannot be defined in time, and that even spatially can recognize different layers' levels of intensity. A crisis may not be evident, and it demands analysis to be recognized. [23]

A definition that captures many of the features of crisis is reported in Canyon [19, p. 8]:

'a damaging event or series of events, that display emergent properties which exceed an organization's abilities to cope with the task demands that it generates and has implications that can effect a considerable proportion of the organization as well as other bodies. The damage that can be caused can be physical, financial, or reputational in its scope. In addition, crises will have both a spatial and temporal dimension and will invariably occur within a sense of 'place'. Crises will normally be 'triggered' by an incident or another set of circumstances (these can be internal or external to the organisation), that exposes the inherent vulnerability that has been embedded within the 'system' over time'.

Weick and Sutcliffe [1] argue that 'unexpected events often audit our resilience. Everything that was left unprepared becomes a complex problem'. Challenging the mindset that purports a 'linear' cause and effect paradigm, Boin et al. [21] regards crisis emerging from '…the result of multiple causes, which interact over time to produce a threat with devastating potential'. Take for example such disaster and crisis events as Deepwater Horizon, Fukushima Daiichi nuclear accident, Ebola outbreak, these events all contained 'resident pathogens' Reason [25] that lied dormant before being triggered by an event (see Masys [7–14], for a disaster forensics perspective). Such disaster events typify complex disasters that require a systems level analysis to reveal the interdependencies, interconnectivity and 'resident pathogens' that permeate the system.

As described in Boin et al. [21]

'The causes of crises thus seem to reside within the system: the causes typically remain unnoticed, or key policy makers fail to attend to them. In the process leading up to a crisis, seemingly innocent factors combine and transform into disruptive forces that come to represent an undeniable threat to the system. These factors are sometimes referred to as pathogens, as they are typically present long before the crisis becomes manifest'.

Understanding these underlying conditions that precipitate disasters have been presented in terms of Normal Accident Theory [24], Resident pathogens, Latent risks and Swiss Cheese Model [24], Normalization of Deviance [26], Incubation [27]. In short, they described non-linear dynamics and complexity of the problem space (system). Boin et al. [21] describes crisis in terms of a complex system with inherent vulnerabilities. Boin et al. [21, p. 7] argues that '…Growing vulnerabilities go unrecognized and ineffective attempts to deal with seemingly minor disturbances continue. The system thus "fuels" the lurking crisis. Only a minor "trigger" is needed to initiate a destructive cycle of escalation, which may then rapidly spread throughout the system. Crises may have their roots far away (in a geographical sense) but rapidly snowball through the global networks, jumping from one system to another, gathering destructive potential along the way'. Such complex disaster events are exemplified by Fukushima Daiichi nuclear accident and 2003 US/Canada Blackout [6] and the current COVID-19 pandemic (described in detail in this volume). The transborder threat associated with COVID-19 and the failing national and global societal systems exemplifies this definition. This further supports the notion of the 'creeping crisis' described in Boin et al. [3].

3 Sensemaking

Deep uncertainty is certainly a characteristic of a crisis and is exemplified in the COVID-19 pandemic. This inherent uncertainty has shaped crisis management and challenged crisis leadership. We once again return to Weick and Sutcliffe [1] statement: 'unexpected events often audit our resilience. Everything that was left unprepared becomes a complex problem'. How is it that through hindsight knowledge we are able to foresee the impending crisis/disaster? Both 9/11 and Fukushima Daiichi disaster point to a '...lack of imagination and a mindset that dismissed possibilities' [6]. The question emerges, how can we better mitigate, prepare for, respond to and recover from disaster events? One of the five key tasks of crisis leadership is 'sensemaking' [21].

Padan [28, p. 219] argues that in today's complex and dynamic world, sensemaking is perceived as a core leadership capability [29]. Weick [30] refers to sensemaking in terms of '...how we structure the unknown so as to be able to act in it. Sensemaking involves coming up with a plausible understanding—a map—of a shifting world; testing this map with others through data collection, action, and conversation; and then refining, or abandoning, the map depending on how credible it is' (cited in [29]). As noted in Macrae [31], the failure of sensemaking is strongly implicated in many studies of organizational accidents and disaster (e.g. [27, 32, 33]). Cited in Stern [34, p. 48], Alberts and Hayes [35, p. 102] argue that '...sensemaking is much more than sharing information and identifying patterns. It goes beyond what is happening and what may happen to what can be done about it'. Sensemaking is essentially about creating an emerging picture of the crisis, developed through data collection, analysis, experience and dialogue across a diverse network of stakeholders and participants. The response to the Boston Marathon attack of 2013 certainly exemplifies this as described in detail in Marcus et al. [36].

The complex problem associated with crisis events (such as the COVID-19 pandemic) requires a framework to support sensemaking to reveal the vulnerabilities and explore the possibility/plausibility space in support of strategic decision making. Systems thinking figures prominently in the discourse pertaining to disaster aetiology and disaster forensics [10, 11]. Jackson [37, p. 65] defines systems thinking paradigm as'...a discipline for seeing the 'structures' that underlie complex situations, and for discerning high from low leverage change...Ultimately, it simplifies life by helping us to see the deeper patterns lying beneath the events and the details'. Systems thinking thereby is a key sensemaking approach to view inherent vulnerabilities in societal systems that lie dormant, waiting for a trigger event to unleash a cascading disaster across communities. There are many systems thinking approaches that can be leveraged to support sensemaking in crisis management: Rich Pictures [38], Cynefin [39, 40], modeling and simulation (such as system dynamics) [41]. The selection of tools and approaches is context specific with the nature of the crisis.

Some key lessons pertaining to COVID-19 crisis management and sensemaking are emerging. Boin et al. (42:192) describes these lessons to support sensemaking:

- Prepare to collect more rather than less data.
- Spread the net widely.
- Expert assessments cannot replace empirical fact-finding efforts.
- Model builders must clearly explicate starting assumptions and limitations of their models.

These lessons are reflected within the crisis leadership framework discussed in the next section.

4 Discussion

Crises conditions associated with volatility, uncertainty, complexity, ambiguity (VUCA) represent potentially high-impact events for which sensemaking is critical. Various regional and global disasters have stress tested the crisis management readiness of governments and organizations. Crisis management is the systematic attempt to identify and detect possible crises and to take actions and measures to prevent them, contain their effects or disruption, and finally recover [43]. Examples of such disaster events include 9/11, Fukushima Daiichi Nuclear accident, Deepwater Horizon Oil disaster, Hurricane Katrina, Ebola outbreak in West Africa, 2003 US/Canada Blackout, global financial crisis of 2008 and public health disasters such as H1N1, H5N1, SARS and now the COVID-19 pandemic. The failure of foresight [27] to inform mitigation, preparedness, response and recovery in these events emerges as a key contributor to the crisis event. This is essentially a sensemaking issue (central to crisis leadership). As discussed in Stern [34, p. 48], '…leaders face recurring challenges when confronted with crisis across the areas of preparing, sensemaking, decision making, meaning making, terminating and learning'.

Research has suggested that inadequate sensemaking can exacerbate crisis conditions and lead to catastrophic outcomes [44]. Crayne et al. [44] reports on the implication of ineffective sensemaking on the proliferation of the West Nile virus. Crayne et al. [44] argues that:

'…akin to the present issue of COVID-19, Weick [45] noted that challenges in sensemaking distributed across multiple institutions were directly related to the Centers for Disease Control's initial misdiagnosis of the West Nile virus during its spread through New York City in the late 1990s. The errors made during this time and an inability to develop comprehensive understanding of the issue resulted, in Weick's [45] view, in the proliferation of a virus that eventually infected and caused harm to millions of Americans. The preponderance of evidence suggests that sensemaking is an essential element to successful navigation of crisis events…and that those who take ownership of that process have significant and direct influence over the success of any crisis response'.

Failed sensemaking Weick [30] has been implicated in crisis leadership disasters. As reported in Nyenswah et al. [17], the Ebola epidemic of 2014–2015 illustrated how ineffective sensemaking was implicated into the disaster aetiology. Other leadership failures are well captured in the reports of such disaster events as 9/11, Hurricane Katrina, and the Fukushima Daiichi Nuclear accident and emerge across the disaster management cycle (prevention, preparedness, response, recovery). These leadership failures make up part of the disaster/crisis aetiology [6], (Masys 2014). As described in Boin et al. [18, p. 41], '…leaders must organize a set of activities that includes **sensemaking (understanding the crisis),** critical decision making, vertical and horizontal coordination, meaning-making (formulating an authoritative definition of the situation), and communication'. Failure in this capacity and capability for sensemaking directly affects the outcomes of crisis management. With this in mind, Ramos et al. [46, p. 3] reports on the value of systems thinking. 'Systems thinking also helps us to identify and understand critical linkages, synergies and trade-offs between issues generally treated separately, and thus to reduce unintended consequences of policies'.

5 Crisis Leadership Framework

As shown in Fig. 1, an analysis of the leadership failures associated with Hurricane Katrina, BP Deepwater Horizon, Ebola epidemic (2014–2016) and Fukushima Daiichi nuclear accident was conducted. Through the application of Grounded Theory [47], key themes emerged pertaining to the leadership failures. An appreciative inquiry lens was then applied (likened to a polarizing lens) to view the failures and map the opposite characteristics thereby creating emergent qualities of high

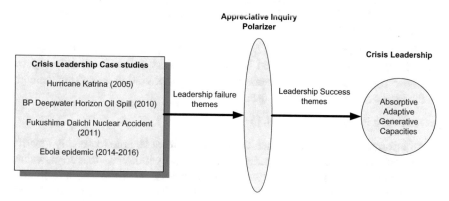

Fig. 1 Development of crisis Leadership framework

performing crisis leadership. The appreciative inquiry lens shifts the paradigm of analysis from a deficit based approach (focusing on what went wrong) to a strength based approach (what are the qualities of outstanding leadership in crisis). As noted in Wattie and Masys [48], 'As a high engagement, strength-based approach to organizational change, Appreciative Inquiry focuses on aligning strengths of the organization with opportunities, aspirations and desired results and transforming goals into action fostering organizational learning at its core'. These were then analyzed through various leadership models to reveal a crisis leadership framework.

The Leadership success themes that emerged from the analysis included:

- Foresight
- Communication
- Growth mindset
- Innovation
- Social interaction
- Adaptation
- Purpose driven
- Reflective learning
- Decision making in ambiguity.

A review of the leadership models in the literature revealed that these leadership success themes mapped to a Generative capacity defined in Surie and Hazy [49]:

> *Generative leadership*, defined as those aspects of leadership that foster **innovation**, organizational **adaptation**, and **high performance** *over time*. A critical element of generative leadership is the ability to seek out, foster, and sustain generative **relationships** [50] that yield **new learning** relevant for innovation. This, in turn, requires a nuanced understanding of the environment and an ability to structure situations and manage interactions.

> Further Surie and Hazy [49] argue that 'Generative leadership is particularly relevant in situations of complexity when uncertainty and rapid change are dominant; since outcomes are not certain, the focus must be on the process. … generative leaders promote **information flow** and **feedback seeking**. A consequence is that exploitation and exploration can be managed concurrently. Similarly, exercising generative leadership involves **problem solving** and **innovation** and suggests that these can be distributed on a wider scale rather than limited to a few organization members at the top of the hierarchy or within a specialized group' (my emphasis in the above quote).

Effective leadership is shaped by the challenges presented and how they are met. Crisis situations characterized by VUCA calls for unique leadership capacities: absorptive, adaptive, and generative [5]. The right kind of leadership depends on the kind of challenges leaders face.

As described in Castillo and Trinh [5]:

- As information flows ever faster, organizations must be able to quickly identify relevant data, assimilate that information, and apply it in ways that create value. This ability, known as **absorptive capacity**, enables a firm to dynamically and continuously innovate … Absorptive capacity is co-created and arises through interactions at multiple levels,… Meaning making is another fundamental aspect of absorptive capacity…

- **Adaptive capacity** loosely refers to the ability of leaders to change to become more fit with the environment in which they operate, including but not limited to modifying existing procedures, adjusting to new circumstances, and updating knowledge and skills to meet new situational demands. In these VUCA times, adaptive capacity is critical at multiple levels of organization.
- While absorptive and adaptive capacities are important to tackle such problems, they alone are not enough. Organizations must also build **generative capacity**, the two-pronged ability to both innovate and evoke affordances that promote ongoing innovation

With the backdrop of the ongoing COVID-19 pandemic, these three capacities certainly resonate with managing a crisis under VUCA conditions, where information and knowledge is emerging and impacts are surprising.

Busche [51] reports on the application of systems thinking (CYNEFIN) in support of problem framing as a sensemaking approach. As noted in Busche [51] research has argued that '… the single most common failure of leadership was to treat adaptive challenges like technical problems. In complex situations, however, a different, *generative change* approach is appropriate. Essentially, generative change requires identifying the issue or problem that needs to be addressed and framing it in a way that will motivate the variety of stakeholders who are "part of the problem" to engage in coming up with new ideas'. Further Busche [51] argues '…When problems are too complex for anyone to analyze all the variables and know the correct answer in advance, the best approach is to use generative change processes to develop adaptive ideas and solutions'. This in essence creates a space for innovation and learning. This is particularly relevant to the COVID-19 pandemic as new vulnerabilities and impacts emerge nationally and globally requiring creative generative leadership.

In short, crisis conditions call for unprecedented levels of innovation. Generative capacity recognizes and embraces this to support an Ecology of Innovation. As cited in Goldstein et al. [52, p. 118] 'Every good manager recognizes the positive role that heterogeneity can play, especially in project groups and cross-functional teams. Generative leadership pushes this view farther, by recognizing the importance of how diversity—across informational differences—generates seeds of novelty leading to highly innovative organizational ecologies. Diversity in this sense does not just mean a broad range of demographic differences but includes differences of perception, difference of mental models, difference in perspectives. …Generative leadership can spark an ecology of innovation by highlighting differences, especially giving voice to those who do not represent the dominant majority'. This diversity is critical in shaping sensemaking and decision making.

Figure 2 consolidates the insights derived from the literature and analysis to put forth a crisis leadership framework that has sensemaking [21, 30] as a central element connecting with the essential components of communication, innovation and engagement. All this is circumscribed and integrated with an absorptive, adaptive and generative capacity (as described in Castillo and Trinh [5]).

This crisis leadership framework certainly resonates with the lessons learned from the current COVID-19 pandemic. As described earlier, the lessons learned

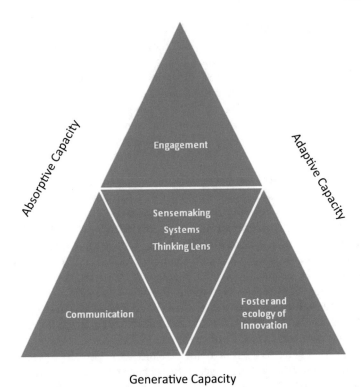

Fig. 2 Crisis leadership framework (absorptive, adaptive, generative capacities) to support decision making under VUCA conditions

pertaining to the COVID-19 crisis management (Boin et al. [42]:192), [53] emerge from the central component of sensemaking and apply to communication, innovation and engagement. The absorptive, adaptive and generative capacities are apropos in managing the 'creeping crisis' [3].

Failure of sensemaking as a central component has resulted in issues pertaining to risk communication, public and stakeholder engagement, and innovation. This is reflected and validated in the findings of [54–59].

6   Operationalizing Crisis Leadership

Successful organizations that can manage complexity and avoid disasters though operating in hazardous and challenging conditions are called High Reliable Organizations (HRO). As described in Masys [13], the roots of the HRO paradigm finds itself in the examination of aircraft carrier operations, air traffic control and nuclear power operations. They not only avoid dire consequences by catching problems early, but they have actually proven over time to have disproportionately fewer problems. These organizations do this by consistently noticing the unexpected, reporting it in an honest way, responding quickly and appropriately, learning from the things they did, and improving the process for the next time a challenge arises.

Weick and Sutcliffe [1] suggest there are five common concepts that help organizations manage the threat of failure, absorb damage and surprises, and thereby become an HRO. As shown in Fig. 3, the crisis leadership framework is shown at the heart of decision making and action. Here we take the framework and operationalize it within the 5 principles of an HRO.

The principles of HRO have been reported and validated in the context of COVID-19 pandemic. As described in Thull-Freedman et al. [60] 'By incorporating principles of high reliability organizing to our pandemic fight, we can lessen the harm that occurs to ourselves and our patients over the coming months and be more prepared to keep ourselves safe during the future challenges we encounter in our always changing environments'. Similarly lessons learned stemming from communication pertaining to the pandemic and HRO principles is covered in Sanders [57]. Therefore operationalizing the crisis leadership framework within the HRO principles is apropos. Within this construct (Framework and HRO) we can include the ten executive crisis tasks described in Boin et al. [18] (Table 1).

Fig. 3 Operationlization of Crisis leadership framework within the HRO principles [1] to support crisis management

Table 1 Executive tasks of crisis management [18]

Task #1: Early recognition
Task #2: Sensemaking
Task #3: Making critical decisions
Task #4: Orchestrating vertical and horizontal coordination
Task #5: Coupling and decoupling
Task #6: Meaning making
Task #7: Communication
Task #8: Rendering accountability
Task #9: Learning
Task #10: Enhancing resilience

It is not enough to purport the qualities of crisis leadership without connecting it to an operational framework that encompasses the executive tasks of crisis leadership. Here, the HRO principles facilitate that operational implementation of the framework for success in crisis situations.

7 Conclusion

Wisittigars et al. [61] argue that crisis management and leadership are interconnected organizational competencies. The COVID-19 pandemic has certainly stress-tested societal systems and global health security. As described in OECD [4, p.],'…the pandemic has reminded us bluntly of the fragility of some of our most basic human-made systems'. By drawing upon recent disaster events and more recently the COVID-19 pandemic, this chapter examined sensemaking within the context of crisis leadership and presented a crisis leadership framework focused on absorptive, adaptive and generative capacities [5] to support problem framing, solution navigation and innovation. As noted in Boin et al. [21], sensemaking emerges as a key leadership capability for the complex and dynamic world we live in today. Leveraging the thought leadership from Weick [30] on sensemaking, a crisis leadership framework was developed and operationalized through the HRO principles. The experiences emerging from the pandemic and captured in the literature all point to the essential element of sensemaking and HRO principles to support crisis management and leadership.

References

1. Weick KE, Sutcliffe KM (2007) Managing the unexpected: resilient performance in an age of uncertainty, 2nd edn. Wiley, San Francisco

2. Taleb NN (2007) The black swan: the impact of the highly improbable. Penguin Books Ltd., London
3. Boin A, Ekengren M, Rhinard M (2020c) Hiding in plain sight: conceptualizing the creeping crisis. Risk Hazards Crisis Public Policy 11(2)
4. OECD (2020) A systemic resilience approach to dealing with Covid-19 and future shocks New Approaches to Economic Challenges (NAEC) 28 April 2020
5. Castillo EA, Trinh MP (2019) Catalyzing capacity: absorptive, adaptive, and generative leadership. J Organ Change Manage 32(3):356–376
6. Masys AJ, Ray-Bennett N, Shiroshita H, Jackson P (2014) High impact/low frequency extreme events: enabling reflection and resilience in a hyper-connected world. In: 4th international conference on building resilience, 8–11 Sept 2014, Salford Quays, United Kingdom. Proc Econ Finan 18:772–779
7. Masys AJ (ed) (2014) Networks and network analysis for defence and security. Springer Publishing
8. Masys AJ (ed) (2014) Disaster management: enabling resilience. Springer Publishing
9. Masys AJ (ed) (2015) Applications of systems thinking and soft operations research in managing complexity. Springer Publishing
10. Masys AJ (ed) (2016) Exploring the Security Landscape: non-traditional security challenges. Springer Publishing
11. Masys AJ (ed) (2016) Disaster forensics: understanding root cause and complex causality. Springer Publishing.
12. Masys AJ (ed) (2018) Security by design. Springer Publishing
13. Masys AJ (2018b) Designing high reliability security organizations for the homeland security enterprise. In: Siedschlag A, Jerkovic A (eds) Homeland security cultures: enhancing values while fostering resilience. Rowman & Littlefield International
14. Masys AJ, Izurieta R, Reina M (ed) (2019) Global health security: recognizing vulnerabilities, creating opportunities. Springer Publishing
15. Taleb NN (2014) Antifragile: things that gain from disorder. Random House, New York
16. Masys AJ (2010) Fratricide in air operations: opening the black-box: revealing the social. PhD Dissertation, June 2010, University of Leicester, UK
17. Nyenswah T, Engineer CY, Peters DH (2016) Leadership in times of crisis: the example of Ebola virus disease in Liberia. Health Syst Reform 2(3):194–207
18. Boin A, Renaud C (2013) Orchestrating joint sensemaking across government levels: challenges and requirements for crisis leadership. J Leadersh Stud 7(3):41–46
19. Canyon D (2020) Definitions in crisis management and crisis leadership. Security Nexus 24:1–10. https://apcss.org/nexus_articles/definitions-in-crisis-management-and-crisis-leadership/
20. Forman R, Atun A, McKee M, Mossialos E (2020) Lessons learned from the management of the coronavirus pandemic. Health Policy 124(2020):577–580
21. Boin A, 't Hart P, Stern E, Sundelius B (2005) The politics of crisis management public leadership under pressure. Cambridge University Press
22. DOD Dictionary of Terms (2020) https://www.jcs.mil/Portals/36/Documents/Doctrine/pubs/dictionary.pdf
23. WHO (2020) https://www.who.int/hac/about/definitions/en/#:~:text=to%20one%20country.-,Crisis,demands%20analysis%20to%20be%20recognized
24. Perrow C (1984) Normal accidents: living with high-risk technologies. Basic Books Inc., New York, NY
25. Reason J (1997) Managing the risk of organizational accidents. Ashgate Publishing, Aldershot, England
26. Vaughan D (1996) The challenger launch decision: risky technology, culture and deviance at NASA. Chicago University Press, London
27. Turner B (1978) Man-made disasters. Wykeham, London
28. Padan C (2017) Constructing crisis events in Military contexts—an Israeli perspective. In: Holenweger M, Jager MK, Kernic F (2017) Leadership in extreme situations. Springer Publishing

29. Ancona DL (2011) Sensemaking framing and acting in the unknown in Handbook of Teaching Leadership, 3–20. Sage Publications Inc, Thousand Oaks, CA
30. Weick KE (1995) Sensemaking in organizations. Sage, Thousand Oaks, CA
31. Macrae C (2007) Interrogating the unknown: risk analysis and sensemaking in airline safety oversight LSE discussion paper no: 43 Date: May 2007. https://eprints.lse.ac.uk/36123/1/Dis spaper43.pdf
32. Snook SA (2000) Friendly fire: the accidental shootdown of US Black Hawks over Northern Iraq. Princeton University Press, Oxford
33. Weick KE (1993) The collapse of sensemaking in organizations: the Mann Gulch disaster. Adm Soc 38(4):628–652
34. Stern EK (2017) Crisis, leadership and extreme contexts. In Holenweger M, Jager MK, Kernic F (eds) Leadership in extreme situations. Springer Publishing.
35. Alberts DS, Hayes RE (2003) Power to the edge. Command and Control Reseearch Program, Washington DC
36. Marcus LJ, McNulty E, Dorn BC, Goralnick E (2014) Crisis Meta-leadership lessons from the Boston Marathon bombing response: the ingenuity of swarm intelligence. https://cdn1.sph.harvard.edu/wp-content/uploads/sites/2443/2016/09/Marathon-Bom bing-Leadership-Response-Report.pdf
37. Jackson MC (2003) Systems thinking: creative holism for managers. Wiley, West Sussex
38. Checkland P (1993) Systems thinking, systems practice: a 30 year retrospective. John Wiley and Sons, New York
39. Snowden DJ, Boone ME (2007) A Leader's framework for decision making. Harvard Bus Rev 85(11):68–76
40. Stikeleather J, Masys AJ (2019) Global health innovation. In Masys AJ, Izurieta R, Reina M (eds) Global health security: recognizing vulnerabilities, creating opportunities. Springer Publishing
41. Taylor I, Masys AJ (2012) A system dynamics model of COVID-19 in Canada: a case study in sensemaking. In: Masys AJ (ed) Sensemaking for security. Springer Publishing
42. Boin A, Lodge M, Luesink M (2020) Learning from the COVID-19 crisis: an initial analysis of national responses. Policy Des Pract 3(3):189–204
43. Constantinides P (2013) The failure of foresight in crisis management: a secondary analysis of the Mari disaster Panos. Technol Forecast Soc Chang 80(2013):1657–1673
44. Crayne MP, Medeiros KE (2020) Making sense of crisis: charismatic, ideological, and pragmatic leadership in response to COVID-19. Am Psychol. Advance online publication
45. Weick KE (ed) (2005) Managing the unexpected: complexity as distributed sensemaking. Uncertainty and surprise in complex systems. Springer, Berlin, Germany, pp 51–65
46. Ramos G, Hynes W, Müller J-M, Lees M (ed) (2019) Systemic thinking for policy making the potential of systems analysis for addressing global policy challenges in the 21st century. OECD SG/NAEC(2019)4
47. Charmaz K (2006) Constructing grounded theory: a practical guide through qualitative analysis. Thousand Oaks
48. Wattie J, Masys AJ (2014) Enabling resilience: an examination of high reliability organizations and safety culture through the lens of appreciative inquiry. In: Masys AJ (ed) Disaster management: enabling resilience. Springer Publishing.
49. Surie G, Hazy JK (2013) Generative leadership: nurturing innovation in complex systems. E:CO 8(4):13–26
50. Lane D, Maxfield R (1996) Strategy under complexity: fostering generative relationships. Long Range Plan 29(April):215–231
51. Busche GR (2019) Generative leadership. Can J Physician Leadersh 3:141–147
52. Goldstein J, Hazy JK, Lichtenstein BB (2010) Complexity and the nexus of leadership: leveraging nonlinear science to create ecologies of innovation. Palgrave Macmillan, New York
53. Boin A, Brock K, Craft J, Halligan J, 't Hart P, Roy J, Tellier G, Turnbull L (2020b) Beyond COVID-19: five commentaries on expert knowledge, executive action, and accountability in governance and public administration.Can Public Adm 63(3)

54. Gurr D, Drysdale L (2020) Leadership for challenging times. ISEA 48(1):2020
55. Jahagirdar S, Chatterjee A, Behera S, Mohapatra A (2020) Response to the COVID-19 pandemic in India: case studies on leadership in crisis situations. Int J Health Allied Sci 9(Supplement 1)
56. Kaul V, Shah VH, El-Serag H (2020) Leadership during crisis: lessons and applications from the COVID-19 pandemic. Gastroenterology 159:809–812
57. Sanders KB (2020) British government communication during the 2020 COVID-19 pandemic: learning from high reliability organizations. Church Commun Cult 5(3):356–377
58. Zhang J, Xie C, Wang J, Morrison AM, Coca-Stefaniak JA (2020) Responding to a major global crisis: the effects of hotel safety leadership on employee safety behavior during COVID-19. Int J Contemp Hospitality Manage 32(11):3365–3389
59. Clark-Ginsberg A, Petrun Sayers EL (2020) (2020) Communication missteps during COVID-19 hurt those already most at risk. J Contingencies Crisis Manage 28:482–484
60. Thull-Freedman J, Mondoux S, Stang A, Chartier LB (2020) Going to the COVID-19 Gemba: using observation and high reliability strategies to achieve safety in a time of crisis. CJEM 2020:1–4. https://www.ncbi.nlm.nih.gov/pmc/articles/PMC7211801/pdf/S14818 03520003802a.pdf
61. Wisittigars B, Siengthai S (2019) Crisis leadership competencies: the facility management sector in Thailand. Facilities 37(13/14):881–896
62. Boin A, Kuipers S, Overdijk W (2013) Leadership in times of crisis: a framework for assessment. Int Rev Public Adm 18(1)

Printed in the United States
by Baker & Taylor Publisher Services